AF753235

CHEMISTRY RESEARCH AND APPLICATIONS

A CLOSER LOOK AT KINASE INHIBITION

Chemistry Research and Applications

Additional books and e-books in this series can be found on Nova's website under the Series tab.

CHEMISTRY RESEARCH AND APPLICATIONS

A CLOSER LOOK AT KINASE INHIBITION

ALEX TOMPSON
EDITOR

nova
science publishers
New York

Copyright © 2020 by Nova Science Publishers, Inc.

All rights reserved. No part of this book may be reproduced, stored in a retrieval system or transmitted in any form or by any means: electronic, electrostatic, magnetic, tape, mechanical photocopying, recording or otherwise without the written permission of the Publisher.

We have partnered with Copyright Clearance Center to make it easy for you to obtain permissions to reuse content from this publication. Simply navigate to this publication's page on Nova's website and locate the "Get Permission" button below the title description. This button is linked directly to the title's permission page on copyright.com. Alternatively, you can visit copyright.com and search by title, ISBN, or ISSN.

For further questions about using the service on copyright.com, please contact:
Copyright Clearance Center
Phone: +1-(978) 750-8400 Fax: +1-(978) 750-4470 E-mail: info@copyright.com.

NOTICE TO THE READER

The Publisher has taken reasonable care in the preparation of this book, but makes no expressed or implied warranty of any kind and assumes no responsibility for any errors or omissions. No liability is assumed for incidental or consequential damages in connection with or arising out of information contained in this book. The Publisher shall not be liable for any special, consequential, or exemplary damages resulting, in whole or in part, from the readers' use of, or reliance upon, this material. Any parts of this book based on government reports are so indicated and copyright is claimed for those parts to the extent applicable to compilations of such works.

Independent verification should be sought for any data, advice or recommendations contained in this book. In addition, no responsibility is assumed by the Publisher for any injury and/or damage to persons or property arising from any methods, products, instructions, ideas or otherwise contained in this publication.

This publication is designed to provide accurate and authoritative information with regard to the subject matter covered herein. It is sold with the clear understanding that the Publisher is not engaged in rendering legal or any other professional services. If legal or any other expert assistance is required, the services of a competent person should be sought. FROM A DECLARATION OF PARTICIPANTS JOINTLY ADOPTED BY A COMMITTEE OF THE AMERICAN BAR ASSOCIATION AND A COMMITTEE OF PUBLISHERS.

Additional color graphics may be available in the e-book version of this book.

Library of Congress Cataloging-in-Publication Data

ISBN: 978-1-53616-973-7

Published by Nova Science Publishers, Inc. † New York

CONTENTS

PREFACE

A Closer Look at Kinase Inhibition opens with a discussion on the roles of various tyrosine kinases inhibitors and molecular mechanisms of acquired resistance to tyrosine kinase inhibitor drugs therapy in colon-rectal cancer.

The authors summarize host cell and viral protein kinases, the progress in their use, and the repurposing of approved kinase inhibitors as broad-spectrum antivirals.

The key kinases, namely CLK1, DYRK1A, CDK5, GSK3 and CK1, are analyzed as data indicate that targeting these proteins may have therapeutic promise against Alzheimer's disease.

Chapter 1 - The introduction in the clinical practice of the Tyrosine Kinase Inhibitors (TKIs) in Chronic myeloid leukemia (CML) has really changed the prognosis of CML patients, offering them long-term survival Despite important advances in the treatment of Colon Rectal Cancer (CRC) with TKIs, it remains one of the deadliest cancer, with an increasing incidence in non-western countries. TKI drugs are efficacious for the targeted therapy of various cancers.

In recent years, many scientists and clinicians have studied the structure and function of tyrosine kinases inhibitors (TKIs) and found out the role of TKIs is vital mediator of signal transduction pathways that regulate cancer cell proliferation, angiogenesis and apoptosis and therefore, have become promising targets for anticancer treatments.

Here, the authors provide an understanding on the role of various tyrosine kinases inhibitors (TKIs) and molecular mechanisms of acquired resistance to TKIs drugs therapy in colon-rectal cancer. This information is very helpful for understanding new hypotheses on drug development to put forwards of new kind of anticancer drugs and to design better therapies to overcome TKIs resistance.

Chapter 2 - Kinases are a part of many cell processes and have been linked to some cancer treatment targets. Protein kinases represent central and multifunctional regulators of a balanced virus-host interaction. Host cell kinases are important for the replication of a number of hemorrhagic fever viruses. Cyclin-dependent kinases (CDKs) are required for replication of viruses that replicate only in dividing cells, non-dividing cells and plays crucial regulatory role in cell cycle, transcription, and replication of many viruses. Several kinase inhibitors developed as anti-cancer drugs, also possess ability to disrupt productive replication and dissemination in viral infected cells and thus exhibit anti-viral activity. Some of the kinase inhibitors (ex. LDC4297, OSU-03012, WV970and BIBX 1382) have demonstrated inhibition of different viruses.

In addition to this, some studies highlight the multiple unexpected ways that viruses use host cell kinases during infection. Consequently, kinase inhibitors are attractive candidates for "repurposing" as broad-spectrum anti-viral agents.This chapter summarizes host cell and viral protein kinases, progress in the use and repurposing of approved kinase inhibitors as broad-spectrum antivirals.

Chapter 3 - Ubiquitously expressed, kinases are key enzymes that regulate the cellular metabolism by catalyzing the phosphorylation of serine/threonine/tyrosine residues by transferring a □-phosphate moiety from ATP. The dysregulation or an imbalance in kinase signaling pathways can produce pathological changes that result in disease. For example, Alzheimer's (AD), the most common neurodegenerative disease, is usually caused by hyper-phosphorylation of Tau kinases. Tau is a microtubule-associated protein that is significantly expressed in healthy neurons and plays a significant role in the neuronal cytoskeleton stabilization. Structurally, the Tau protein is divided into four domains: an acidic region in the N-terminal part, a proline-rich region, a microtubule-binding domain (MBD) containing four repeat units and a C-terminal region respectively. Interestingly, each repeat domain in the MBD contains a conserved consensus motif KXGS, which is readily phosphorylated, resulting in the destabilization of the neuronal cytoskeleton. Consequently, phosphorylation disrupts Tau-dependent cellular functions, thereby promoting Tau dissociation from microtubules, forming neurofibrillary tangles (NFT), a pathological hallmark of AD. Recent studies have shed light on the pathological phosphorylation sites responsible for Tau aggregation. This information has led to research involving the development of inhibitors targeting these kinases. In AD, Tau phosphorylation precedes Tau aggregation and the subsequent formation of NFT. Several families of kinases are known to be involved and are well documented and reviewed elsewhere. This chapter focuses on the key kinases, namely CLK1, DYRK1A, CDK5, GSK3□ and CK1□ as data indicate that targeting these proteins may have therapeutic promise against AD. Moreover o*ptimization techniques,* including chemical modifications of small heterocyclic nucleus, that are *currently being used for* the development of highly potent, safe and selective inhibitors has been discussed here.

In: A Closer Look at Kinase Inhibition
Editor: Alex Tompson
ISBN: 978-1-53616-973-7
© 2020 Nova Science Publishers, Inc.

Chapter 1

TARGETED THERAPY OF TYROSINE KINASE INHIBITOR DRUGS FOR TWO DIFFERENT CANCERS (COLON-RECTAL CANCER AND CHRONIC MYELOGENOUS LEUKEMIA) AND THEIR RESISTANCE MECHANISMS

Gabriella Marfe[1,*], Stefania Perna[1] and Carla Di Stefano[2]

[1]Department of Scienze e Tecnologie Ambientali, Biologiche e Farmaceutiche, University of Campania "Luigi Vanvitelli," Caserta, Italy

[2]Department of Hematology, "Tor Vergata" University, Rome, Italy

ABSTRACT

The introduction in the clinical practice of the Tyrosine Kinase Inhibitors (TKIs) in Chronic myeloid leukemia (CML) has really changed

* Corresponding Author's Email: gabmarfe@alice.it.

the prognosis of CML patients, offering them long-term survival Despite important advances in the treatment of Colon Rectal Cancer (CRC) with TKIs, it remains one of the deadliest cancer, with an increasing incidence in non-western countries. TKI drugs are efficacious for the targeted therapy of various cancers.

In recent years, many scientists and clinicians have studied the structure and function of tyrosine kinases inhibitors (TKIs) and found out the role of TKIs is vital mediator of signal transduction pathways that regulate cancer cell proliferation, angiogenesis and apoptosis and therefore, have become promising targets for anticancer treatments.

Here, we provide an understanding on the role of various tyrosine kinases inhibitors (TKIs) and molecular mechanisms of acquired resistance to TKIs drugs therapy in colon-rectal cancer. This information is very helpful for understanding new hypotheses on drug development to put forwards of new kind of anticancer drugs and to design better therapies to overcome TKIs resistance.

Keywords: targeted therapy, drug resistance, tyrosine kinases inhibitor, colon-rectal cancer, CML

ABBREVIATIONS

4E-BP1	Eukaryotic translation initiation factor 4E (eIF4E)-binding protein 1
AOEs	arterial occlusive events
(AP)-CML	accelerated phase
AKT	protein kinase B
AR	Amphiregulin
(BP)-CML	blast phase
BID	twice a day (bis in die latin)
BRAF	v-Raf murine sarcoma viral oncogene homolog B
CCgR	complete cytogenetic response
(CP)-CML	chronic phase
CR	CRIPTO-1
DLTs	dose-limiting toxicities
eIF4E	eukaryotic initiation factor 4E
FAK	focal adhesion kinase

FOXO1	Forkhead box protein O1
HRAS	Harvey rat sarcoma viral oncogene homolog,
KRAS	Kirsten rat sarcoma viral oncogene
IGF-1R	insulin growth factor receptor I
JAK	Just Another Kinase
MAP	Mitogen-activated protein kinase
Mcg	major cytogenetic responses
MCyR	major cytogenetic response
MDR	Multi drug resistance
MEK	Mitogen-activated protein kinase kinase
MMP-7	metalloproteinase-7
mTOR	mammalian target of rapamycin.
NF-κB	nuclear factor kappa-light-chain-enhancer of activated B cells
NRAS	Neuroblastoma RAS viral oncogene homolog
OV	Overall Survival
ORR	Objective Response Rate
p70S6K1	p70 ribosomal S6 kinase 1
PDGFR	Platelet-derived growth factor
PIK3CA	phosphoinositide-3-kinase, catalytic, alpha polypeptide
PI3K	phosphatidylinositol-3-kinase
PP2A	Protein phosphatase 2A
PRKCH	protein kinase C eta
RAF	Rapidly Accelerated Fibrosarcoma
RAS	Rat sarcoma
RCT	randomized controlled trials
SMAD	an acronym from the fusion of Caenorhabditis elegans Sma genes and the Drosophila Mad, Mothers against decapentaplegic
S1P	Sphingosine-1-phosphate
S1P2	Sphingosine-1-phosphate receptor 2
SphK1	Sphingosine kinase 1
STAT	Signal Transducer and Activator of Transcription

TDM-1	Trastuzumab-emtansine
TGF-α	Transforming Growth Factor
TNM	Tumour, Node, Metastasis

INTRODUCTION

Colorectal cancer (CRC) is a relatively common tumor with high incidence and the lifetime risk in industrialized countries is about 5% on cancers affecting both men and women worldwide [1]. Every year over one million of new cases are reported and around 600,000 patients die from this cancer (Figure 1) [2, 3]. Various reports have shown that the epidermal growth factor receptor (EGFR) is overexpressed in about 50% of patients with CRC [4, 5]. The EGFR gene, located on Chr 7p12–13, encodes for a 170 kDa transmembrane receptor with intrinsic tyrosine kinase (TK) activity belonging to the ErbB family that includes ErbB1 (EGFR or HER1), ErbB2 (HER2), ErbB3, and ErbB4. The binding between these receptors and ligands leads to receptor-dimerization with another EGFR monomer or with a monomer of another ErbB family member and activates the autophosphorylation of different tyrosine residues in the intracellular domain. Such process triggers the activation of the RAS/RAF/MEK/MAPK pathway and induce activation of important tumorigenic processes, including proliferation, apoptosis, angiogenesis, and invasion (Figure 2) [6, 7]. Moreover, several authors have reported the association of activated and transforming RAS genes in different human cancers. Three human RAS genes are identified (H-RAS, N-RAS and K-RAS) and encode proteins that activate several pathways, including the RAF-MEK-ERK/MAPK cascade. The KRAS or BRAF mutations arise in about 50% of the patients with CRC and are associated with poor prognosis [8]. The PI3K/AKT/mTOR pathway is also involved in colon tumorigenesis. Furthermore, many studies have found that phosphatase and tensin homolog (PTEN) (a negative regulator of Akt) is activated in 60-70% of human CRC [9]. Two diverse protein complexes (mTOR complex 1 and mTOR complex 2) form Mammalian target of rapamycin (mTOR) [10, 11]. The mTORC1 signaling pathway is activated by pAkt and two targets of such complex are the serine/threonine kinase p70S6K1 and 4EBP1. The mTORC1 (mTOEc) phosphorylates 4EBP1 at multiple sites and causes the dissociation of eIF4E from 4EBP1, decreasing the inhibitory effect of 4EBP1 on eIF4E-dependent translation initiation [12]. Rapamycin prevents protein translation through inhibition of mTOR and 4EBP1 dephosphorylation [13]. The Food and Drug Administrator (FDA) and European Medicines [14, 15] has been approved two anti-EGFR MoAbs for CRC: Cetuximab was approved as first line treatment for metastatic CRC (mCRC) in combination with 5-Fluorouracil and Irinotecan (FOLFIRI) or Oxaliplatin (FOLFOX) in wild-type KRAS tumors [14-16]. In different preclinical and clinical studies, the authors showed that cetuximab inhibited the EGFR pathway [17, 18]. Other trials showed that cetuximab in combination with irinotecan had a response rate 23% in patients with advanced CRC [19, 20]. Different studies reported that patients with activating mutations in KRAS or BRAF had little or no response to cetuximab therapy [21-24]. A new mechanism of de novo resistance to cetuximab therapy has observed and it happened via increased signalling through both human epidermal growth factor receptor 2 (HER2) gene amplification and increased secretion of the HER2 ligand heregulin. In fact, an increased expression of HER2 protein has been reported in up to 20% of patients with mCRC whereas HER2 gene amplification was a rare event in this disease, occurring in approximately 2% of cases [25]. In addition, a recent report [26] has analyzed with a next-generation sequencing (NGS)-based technique tumor samples from the 'cetuximab after progression in wild-type KRAS colorectal cancer patients' (Capecitabine plus Irinotecan-CAPRI) Gruppo Oncologico dell'Italia Meridionale (GOIM) clinical trial) [27]. Such report has revealed that 30/45 cases with mutated KRAS had concomitant mutations in other genes, with samples showing up to five different mutations. Furthermore, the authors have observed that in some cases, the frequency of mutant alleles did not correlate with the fraction of neoplastic cells. Finally, in tumors with multiple mutations, the frequency of mutant alleles was often different among the affected genes, thus suggesting a further level of intratumor heterogeneity [26]. Subsequently, in another study, the same authors evaluated the efficacy and safety of FOLFIRI plus cetuximab in age-defined subgroups. A post-hoc analysis was performed in CAPRI trial patients: progression-free survival (PFS), overall response rate (ORR) and safety were analysed by age-groups and stratified according to molecular characterisation. 340 patients with mCRC were treated in first line with FOLFIRI plus cetuximab. Among patients 104 were wild- type (wt) for KRAS, NRAS, BRAF, PIK3CA genes. Median PFS and ORR was similar within the age-subgroups [28]. Additionally, in another study, the tumor samples mCRC patients of trhis trial were retrospectively assessed by next generation sequencing (NGS). Both NGS analysis of tumor tissue and plasma testing with magnetics digital PCR (BEAMing). BEAMing identified RAS mutations in 33/92 patients. However, 10 cases were RAS tissue mutant and plasma wild-type, and additional 10 cases were tissue wild-type and plasma mutant, Analysis of plasma samples with droplet digital PCR (ddPCR) or deep-sequencing detected RAS mutations in 2/10 tissue mutant, plasma wild-type patients. In contrast, in all tissue wild-type and plasma mutant cases, ddPCR or deep-sequencing analysis of tumor tissue confirmed the presence of RAS mutations at allelic frequencies This study indicated that RAS testing of liquid biopsy results in a similar outcome when compared with tissue testing in mCRC patients receiving first-line anti-EGFR monoclonal antibodies [29]. In another trial called PETACC8 (cetuximab + FOLFOX vs FOLFOX), the authors analyzed the results of the full RAS and BRAF wt patients with resected stage III colon cancer. They found that 745 patients were known to have KRAS exon 2 mutations while 163 had the BRAF V600E mutation. Of the remaining 1651 pts, 1054 were assessed by NGS, showing that a further 227 patients had KRAS exon 2, 3, 4 or NRAS exon 2, 3, 4 mutations, and that 46 of them had a newly diagnosed BRAF mutation. In the overall trial population, NRAS and KRAS codon 61 mutations were the only rare mutations with the same pejorative prognostic value as KRAS exon 2 or BRAF V600E mutations. The clinically relevant 0.76 adjusted hazard ratio (HR) observed for disease-free survival (DFS) in favour of adding cetuximab to FOLFOX, in full RAS and BRAF wt stage III colon cancer patients, may justify a new randomized controlled trial testing EGFR inhibitor in this setting [30]. A meta-analysis was performed investigate wt

KRAS mCRC after the first-line treatment, and treatment with FOLFOX/FORFIRI±cetuximab. 1464 patients were included in this meta-analysis. Compared, cetuximab as the first-line therapy has improved overall survival (OS) when compared to FOLFOX/FORFIRI. The addition of cetuximab in mCRC as the first-line treatment is a potential effective approach in the improved outcomes but associated with increased toxicity [31]. A multicenter proof of concept phase II trial study was conducted in patients with advanced colorectal cancer receiving FOLFIRI-cetuximab regimens to explore individual drug tailoring using pharmacogenetics and pharmacokinetics (PK) monitoring. Patients were stratified by their pharmacogenetic/phenotypic status for this reason the irinotecan dose was in according to the number of TA tandem repeats in the UGT1A1 promoter, while the 5-fluorouracil (5-FU) dose was initially adjusted according to dihydropyrimidine dehydrogenase (DPD) activity. Median progression-free survival (PFS) for all 85 patients was 181 days and 200, 132, and 121 days for patients with 6/6, 6/7, and 7/7 UGT1A1 genotypes, respectively; these differences were not statistically different. In parallel, a strong relationship was shown between cetuximab AUC and regimen efficacy. The authors concluded that personalized drug tailoring when administering in FOLFIRI + cetuximab allows for safe and efficient individual dose intensification [32]. Another phase II study investigated the efficacy and safety of biweekly cetuximab combined with standard oxaliplatin-based chemotherapy [infusional 5-fluorouracil (5-FU), leucovorin, and oxaliplatin (FOLFOX-6)] in the first-line treatment of KRAS wild-type metastatic colorectal cancer (mCRC). Sixty patients received a biweekly intravenous infusion of cetuximab followed by FOLFOX-6. The median progression-free survival was 13.8 months. Thirteen patients (21.7%) were able to undergo resection of previously unresectable metastases, with the aim of curing them. The median follow-up was 22.7 months, and median overall survival was 31.0 months. Cetuximab did not increase FOLFOX-6 toxicity and was generally well tolerated. The results of this current study demonstrated that the combination of biweekly cetuximab with FOLFOX-6 was well tolerated and had a manageable safety profile for the first-line treatment of KRAS wt metastatic colorectal cancer [33]. Another study aimed to clarify the clinical benefits of this treatment regimen when used as a first-line therapy in patients with expanded RAS/BRAF/PIK3CA wt mCRC, using the data and tumor specimens from two previously published Phase II clinical trials. The gene mutation status and clinical data of 102 patients with KRAS wt mCRC, who received either of Capecitabine Plus Oxaliplatin (CAPOX) + cetuximab or FOLFOX + cetuximab, were analyzed. The primary endpoint was response rate (RR) of CapeOX + cetuximab treatment in patients with extended RAS/BRAF/PIK3CA wt mCRC. The authors identified 88 patients with extended RAS/BRAF/PIK3CA wt mCRC. Those treated with CapeOX + cetuximab had a 61.5% RR (while those treated with FOLFOX + cetuximab had a 66.7% RR. Patients with any mutation (n = 14) had a 42.9% RR. There were no significant differences between these three groups (P = 0.298). The disease control rate was 86.5% in the CapeOX + cetuximab group, and 88.9% in the FOLFOX + cetuximab group. Maximum tumor size changes were largest in patients with wt mCRC treated with FOLFOX + cetuximab followed by patients with wt mCRC treated with CapeOX + cetuximab, and then by those with any mutation. In summary, patients with RAS/BRAF/PIK3CA wt mCRC had a sufficient RR following first-line treatment with CapeOX + cetuximab. These results suggested that this combination therapy should be considered as a treatment option for patients with advanced mCRC [34]. In another trial, the Irinotecan Cetuximab Evaluation and Cetuximab Response Evaluation (ICECREAM) were assessed the efficacy of cetuximab monotherapy compared with cetuximab combined with chemotherapy for quadruple wt (KRAS, NRAS, BRAF, or P13KCA exon 20) metastatic colorectal cancer [35]. In comparison with cetuximab alone, cetuximab plus irinotecan increases the response rate and delays progression in irinotecan-resistant RAS wt colorectal cancer [35]. Another multicentre, randomised, and phase II study evaluated mFOLFOX+cetuximab followed by maintenance (modifed) mFOLFOX+cetuximab or single-agent cetuximab in metastatic colorectal cancer patients (NCT01161316). Previously, untreated mCRC patients (wt KRAS) were randomised to receive cetuximab+mFOLFOX-6 (8 cycles for 2 weeks) followed by maintenance therapy: single-agent cetuximab (Arm-A) or mFOLFOX-6 + cetuximab (Arm-B) until progression. There were no statistically significant differences in the progression-free survival PFS between two arms. This phase II trial suggested that maintenance therapy with single agent cetuximab following mFOLFOX+cetuximab induction could be a valuable option compared with mFOLFOX+cetuximab treatment continuation [36]. In SAKK 41/10 trial, the benefit of cetuximab, either alone or in combination with capecitabine, was evaluated in vulnerable elderly patients with RAS/BRAF-wt mCRC. The trial was stopped prematurely due to slow accrual after the inclusion of 24 patients (11 in the monotherapy arm, 13 in the combination arm). At week 12, 6 of 11 patients were progression-free in the cetuximab monotherapy arm and 9 of 13 patients in the combination arm. Response rate was 9% in the monotherapy arm and 38% combination arm. The 6 patients with right-sided primary tumors were not responsive to cetuximab. NGS revealed additional mutations affecting the RAS/RAF/MAP kinase pathway in 5 patients; 4 of these patients showed early disease progression. In the combination arm, a higher incidence of toxicities and treatment stoppings was observed. In conclusion, trial recruitment - requiring both geriatric as well as molecular eligibility criteria - proved more difficult than expected [37]. Cancer genomics and translational medicine rely on the molecular profiling of patient's tumor obtained during surgery or biopsy. Alternatively, blood is a less invasive source of tumor DNA shed, amongst other ways, as cell-free DNA (cfDNA). Highly sensitive assays capable to detect cancer genetic events from patient's blood plasma became popularly known as liquid biopsy (LqB). Importantly, retrospective studies including small number of selected patients with mCRC patients treated with anti-EGFR therapy have shown that LqB was capable to detect the acquired clonal mutations in RAS genes leading to therapy resistance. However, the usefulness of LqB in the real-life clinical monitoring of these patients still lack additional validation on controlled studies. In this context, the authors designed a prospective LqB clinical trial to monitor newly diagnosed KRAS wt mCRC patients who received a standard FOLFIRI-cetuximab regimen. . A total of 2,178 cell free DNA (cfDNA) mutation analyses were performed and the authors observed that: a) continued wt circulating status was correlated with a prolonged response; b) smoldering increases in mutant (cfDNA) c were correlated with acquired resistance; while c) mutation upsurge/explosion anticipated a remarkable clinical deterioration. The current study provides evidences, obtained for the first time in an unbiased and prospective manner, that reinforces the utility of LqB for monitoring mCRC patients [38]. In the TAILOR trial (ClinicalTrials.gov identifier: NCT012287349) was considered the addition of cetuximab to first-line FOLFOX prospectively choosing a RAS wt population and thus providing confirmative data for the efficacy and safety of cetuximab plus FOLFOX versus FOLFOX alone. In a population of 393 patients with RAS wt mCRC, adding cetuximab to FOLFOX-4 significantly improved the primary end point of progression-free survival time compared with FOLFOX-4 alone. In conclusion, such study met all of its objectives and relevant clinical end points, confirming cetuximab in combination with FOLFOX as an effective standard-of-care first-line treatment regimen for patients with RAS wt mCRC [39]. In other study, the authors investigated the effect of cetuximab retreatment on predictive markers. Between November 2012 and March 2017, 14 patients with KRAS protooncogene GTPase exon 2 wt mCRC who exhibited a clinical benefit (confirmed stable disease for at least 6 months or a clinical response) to an initial cetuximab-based regimen, who received multiple lines of chemotherapy following disease progression and ultimately received a second cetuximab and irinotecan regimen, were analyzed. For retreatment, patients received bi-weekly irinotecan combined with cetuximab Initial cetuximab treatment was administered as a first-line treatment in 11 patients, a second-line treatment in 1 patient and a third-line treatment in 2 patients The objective response rate of cetuximab retreatment was 21.4% and the median PFS was 4.4 months. The Spearman's correlation coefficient for the PFS following retreatment and duration of initial cetuximab-based regimens demonstrated a more marked correlation compared with that between the PFS following retreatment and the interval time between the two regimens. In sommary, cetuximab retreatment may provide clinical benefit to mCRC patients, who were good responders with longer periods of initial cetuximab-based therapy [40]. High miR-31-3p expression is associated with inferior outcomes in KRAS wt advanced colorectal cancer patients treated with anti-EGFR therapy. For this reason, miR-31-3p expression were evaluated in patients with operable colorectal liver metastases (LM) enrolled in the new EPOC study. MiR-31-3p expression was measured in primary tumors (PT) from 149 KRAS wt patients including 71 receiving chemotherapy alone (CT) and 78 receiving chemotherapy plus cetuximab (CTX) MiR-31-3p expression was also measured in LM from 94 patients with tumor tissue available. The median progression-free survival for the combined populations with mid or high miR-31-3p expression was shorter in the CTX versus the CT arm Low miR-31-3p expressers had similar outcomes irrespective of treatment and its expression was correlated between paired PT and LM samples in the CT group but not in the CTX group. Patients with low miR-31-3p expression in the New EPOC study were not harmed by the addition of cetuximab. This supports miR-31-3p as a promising predictive biomarker for anti-EGFR therapy in KRAS wt advanced colorectal cancer [41]. In addition, advanced colorectal cancer (ACRC) patients treated with cetuximab plus chemotherapy at Sun Yat-sen University Cancer Center from January 1, 2005, to October 1, 2015, were studied. Expression of autophagy-related proteins Beclin1, (microtubule-associated protein) 1A/B-light chain 3 (LC3), and 4E-binding protein 1 (4E-BP1)] was examined by Western blot in CRC cells and by immunohistochemistry in cancerous and normal tissues. In 68 ACRC patients, immunohistochemistry showed that Beclin1 levels were significantly correlated with those of LC3 and 4E-BP1 in ACRC tissues. Furthermore, LC3 was significantly overexpressed in tumor tissues compared to normal tissues. In 45 patients with wt KRAS, the expression levels of these three proteins were not related to progression-free survival; however, the expression levels of Beclin1 and 4E-BP1, pathological grade ($P = 0.002$), and T stage were independent prognostic factors for OS. LC3 is overexpressed in tumor tissues, and Beclin1 and 4E-BP1 could be significant predictors of OS in advanced colorectal cancer (ACRC) patients treated with cetuximab [42]. Ledys et al. in another study investigated the effects of anti-MEK (mitogen-activated protein kinase) antibody (trametinib) combined with anti-EGFR (cetuximab) on colon cancer cell lines with different RAS statuses. Even though cetuximab has no effect on RAS cell viability and ERK (extracellular-signal-regulated kinase) phosphorylation (one of the last kinases of the EGFR pathway), trametinib can induce cell death and inhibit the activation of ERK alone or in combination with cetuximab. In a more pathologic context, the authors observed that KRAS colon cancer patient biopsies treated ex vivo with trametinib and cetuximab also presented less ERK phosphorylation. Finally, nine ovarian, endometrial and colon cancer patients with different KRAS statuses were treated with anti-EGFR/anti-MEK combination off label after molecular tumor board decision. KRAS exon 2 patients have significantly longer PFS than with previous lines of treatments. The authors believed that such observations provide a rationale for designing a clinical trial to test this association in RAS exon 2 mutated cancers [43]. Finally, an aberrant activation of the intracellular tyrosine kinase Src has been implicated as a mechanism of acquired chemotherapy resistance in mCRC. For this reason, the oral tyrosine kinase Src inhibitor, dasatinib, was investigated in combination with FOLFOX and cetuximab mCRC patients. The authors performed a phase IB/II study of 77 patients with previously treated mCRC, that received FOLFOX6 with cetuximab and escalating doses of dasatinib (100, 150, 200 mg daily), followed by a 12-patient expansion cohort at 150 mg. Phase II studies evaluated FOLFOX plus dasatinib 100 mg in KRAS c12/13mut patients or in combination with cetuximab if KRAS c12/13 wt FAK and paxillin were utilized as surrogate blood biomarkers of Src inhibition, and paired biopsies of liver metastases were obtained in patients in the expansion cohort. The combination of dasatinib plus FOLFOX with or without cetuximab showed only modest clinical activity in refractory colorectal cancer. This appears to be primarily due to a failure to fully inhibit Src at the achievable doses of dasatinib. The combination of dasatinib plus FOLFOX with or without cetuximab did not show meaningful clinical activity in refractory colorectal cancer due to failure to fully inhibit Src [44].

Panitumumab is a fully human IgG2 monoclonal antibody that is highly selective for the EGFR, which is overexpressed in 25-77% of colorectal cancers and is often associated with a poor prognosis. Binding of panitumumab to EGFR reduces cell proliferation and mediator production and induces apoptosis. In a comparative, phase III trial in adult patients with chemotherapy-refractory metastatic colorectal cancer, intravenous panitumumab 6 mg/kg every 2 weeks plus best supportive care (BSC) improved progression-free survival (PFS) [primary endpoint] and objective tumor response rate to a significantly greater extent than BSC alone. The improvement in PFS produced by panitumumab monotherapy was significantly greater in patients with wt KRAS than in those with mutant KRAS (in whom no benefit from panitumumab was observed). Similarly, all patients experiencing a partial response had wild-type KRAS, while stable disease was achieved by more patients with wild-type KRAS than with mutant KRAS. The predictive value of mutant KRAS for a lack of clinical benefit with panitumumab monotherapy was supported by results from an open-label extension of the phase III study and a large phase II study. Although most patients treated with panitumumab experienced at least one adverse event, the incidence of severe adverse events resulting in discontinuation of treatment was relatively low. The most commonly reported treatment-related adverse events were skin-related toxicities, which reflect the mechanism of action of panitumumab [45]. Like cetuximab, the response to panitumumab can be affected by K-Ras mutations. A phase 3 trial that compared the effects of panitumumab monotherapy with those of best supportive care in patients with metastatic CRC showed that panitumumab was only effective in patients with wild-type K-Ras tumors; response rates were 17% vs 0% in patients with mutant versus wild-type K-Ras tumors [46]. Additionally, the association of panitumumab with FOLFIRI in second-line treatment in wt KAS mCRC patients, significantly improved PFS vs FOLFIRI alone [47]. Two randomized first-line panitumumab trials were analyzed in a study in order to better understand the treatment outcomes by primary tumor sidedness for RAS WT patients: PRIME (phase 3; NCT00364013) compared panitumumab plus FOLFOX versus FOLFOX alone; PEAK (phase 2; NCT00819780) compared panitumumab plus FOLFOX versus bevacizumab plus FOLFOX. Primary tumors located in the cecum to transverse colon were coded as right-sided, while tumors located from the splenic flexure to rectum were considered left-sided. Tumor sidedness ascertainment (RAS WT population) was 83% (n = 559/675); 78% of patients (n = 435) had left-sided and 22% (n = 124) had right-sided tumors. Patients with right-sided tumors did worse for all efficacy parameters compared with patients with left-sided disease in the wt

RAS population and also in the wt RAS/BRAF subgroup. In patients with left-sided tumors, panitumumab provided better outcomes than the comparator treatment, including on median overall survival (PRIME: 30.3 versus 23.6 months, adjusted hazard ratio = 0.73, P = 0.0112; PEAK: 43.4 versus 32.0 months, adjusted hazard ratio = 0.77, P = 0.3125). The results of these retrospective analyses confirmed that in wt RAS patients, right-sided primary tumors were associated with worse prognosis than left-sided tumors, regardless of first-line treatment received. Wt RAS patients with left-sided tumors derived greater benefit from panitumumab-containing treatment than chemotherapy alone or combined with bevacizumab, including an overall survival advantage (treatment difference: PRIME 6.7 months; PEAK 11.4 months). No final conclusions regarding optimal treatment could be drawn for wt RAS WT patients with right-sided mCRC due to the relatively low number of patients. Further research in this field is warranted [48]. In another study, chemotherapy-naive patients with unresectable liver metastases from colon cancer with no other metastatic disease sites were enrolled. All patients received upfront therapy with Capecitabine plus oxaliplatin (XELOX) plus panitumumab (P-XELOX) and were reevaluated for resectability every 4 cycles. The primary endpoint was the objective response rate (ORR). Secondary endpoints were overall survival (OS), progression-free survival, the percentage of patients whose disease became radically resectable and the safety of the P-XELOX combination. Among the 49 patients, 35 presented wild-type KRAS (wt KRAS) while 14 had unknown (9 patients) or mutated (5 patients) KRAS mutational status. Forty-six patients were evaluable for response. After conversion P-XELOX therapy, the ORR in the general population was 54%, with 2 complete responses, 23 partial responses, and 14 cases of stable disease. In patients with wt KRAS, the ORR of the patients reached 65%, which allowed 15 patients with initial unresectable liver metastasis to be reclassified as having resectable disease. Survival analysis demonstrated a median progression-free survival of 8.5 months and a median OS of 21.9 months. Patients who underwent surgery were found to have a significantly better OS when compared with those who did not undergo surgery ($P < .001$). Overall, toxicities were found to be predictable and manageable, with the most common being cutaneous, gastrointestinal, and neurologic toxicities. Conversion P-XELOX therapy yields high response and resectability rates for patients with metastatic colon cancer with extensive liver involvement [49]. A prospective clinical trial was designed to evaluate the efficacy and safety of fluorouracil monotherapy combined with panitumumab administered to patients with KRAS wt mCRC intolerant to oxaliplatin and irinotecan. This single-arm, open-label multicenter phase II trial recruited patients with KRAS wt mCRC from 16 institutes between January 2012 and October 2014. Panitumumab (6 mg/kg) was intravenously administered every 2 weeks, combined with fluorouracil monotherapy, in 2-week cycles. Forty patients met eligibility criteria and received 7 cycles (median) of fluorouracil chemotherapy combined with panitumumab. There were no treatment-related deaths. Median time to treatment failure was 3.2 months. 23 patients experienced at least one adverse effect ≥ grade 3. The response rate was 10.0%. Median progression-free survival and overall survival were 4.3 and 11.3 months, respectively. Total lactase dehydrogenase (LDH) levels and those of LDH-4 and LDH-5, quickly changed with disease reduction or progression. Fluorouracil monotherapy combined with panitumumab was safely administered to patients with KRAS WT mCRC intolerant to oxaliplatin and irinotecan [50]. The primary tumor location has a prognostic impact in mCRC, for this reason, some authors reported the results from retrospective analyses assessing the effect of tumor location on prognosis and efficacy of second- and later-line panitumumab treatment in patients with RAS wt mCRC and on prognosis in all lines of treatment in patients with RAS mutant mCRC. RAS wt data (n = 483) from 2 randomized phase III panitumumab trials (ClinicalTrials.gov identifiers, NCT00339183 and NCT00113763) were analyzed for treatment outcomes stratified by tumor location. The second analysis assessed the effect of tumor location in RAS mutant patients (n = 1205) from 4 panitumumab studies (ClinicalTrials.gov identifiers, NCT00364013, NCT00819780, NCT00339183, and NCT00113763). Primary tumors located in the cecum to transverse colon were coded as right-sided; those located from the splenic flexure to the rectum were coded as left-sided. Of all patients, the tumor location was ascertained for 83% to 88%; 71% to 77%

of patients had left-sided tumors. RAS wt patients with right-sided tumors did worse for all efficacy parameters compared with those with left-sided tumors. The patients with left-sided tumors had better outcomes with panitumumab than with the comparator treatment. Because of the low patient numbers, no conclusions could be drawn for right-sided mCRC. The prognostic effect of tumor location on survival was unclear for RAS mutant patients. These retrospective analyses have confirmed that RAS WT right-sided mCRC is associated with a poor prognosis, regardless of the treatment. RAS wt patients with left-sided tumors benefitted from the addition of panitumumab in second or later treatment lines [51]. In other trial, patients with confirmed metastatic colon or rectum adenocarcinoma, wt KRAS gene tumor exon 2 status, clinical/radiologic disease progression or toxicity during irinotecan or oxaliplatin treatment, and no previous anti-EGFR therapy were randomized 1:1 to receive best supportive care (BSC) with or without panitumumab (6.0 mg/kg, intravenously, on day 1 of each 14-day cycle) in this open-label, multicenter, phase III study (20100007). RAS and BRAF mutation status were determined using Sanger sequencing. Furthermore, early tumor shrinkage (ETS) and depth of response (DpR) were calculated as the percentage change for tumor shrinkage at nadir versus baseline. Overall, 270 patients had RAS wt type mCRC (panitumumab with BSC, n = 142; BSC, n = 128). For patients with wt RAS tumors, median overall survival and progression-free survival were improved with panitumumab with BSC versus BSC. Similar improvements were seen for patients with wt RAS, and wt BRAF tumors. Median DpR was 16.9% for the evaluable panitumumab with BSC wt RAS population. Overall, 69.5% experienced any type of tumor shrinkage at week 8; 38.2% experienced ≥ 20% shrinkage. Similar improvements in OS and PFS were seen with stratification according to ETS. This analysis showed that panitumumab improved outcomes in wild type RAS mCRC and indicated that ETS and DpR could be used as additional efficacy markers [52]. A clinical trial assessed the efficacy and toxicity of panitumumab combined with oxaliplatin and capecitabine as first-line treatment in KRAS exon 2 mCRC patients. Patients with exon 2 KRAS wt mCRC received panitumumab 9 mg/Kg, oxaliplatin 130 mg/m^2, and capecitabine 2000 mg/m^2 repeated every

3 weeks. The primary endpoint was objective response rate (ORR, minimum 42 responses). The authors retrospectively assessed mutations in genes implicated in CRC with massively parallel sequencing; ERBB2 and EGFR amplification with fluorescence in situ hybridization, and tumor-infiltrating lymphocyte density. Among 78 patients enrolled, 45 completed 6 cycles. Objective response occurred in 43 of the patients (complete 6.4% and partial response 48.7%; stable 17.9% and progressive disease 7.7%), while 3.8% were non-evaluable and 15% discontinued their treatment early. Additional mutations in KRAS/NRAS/BRAF were found in 11/62 assessable (18%) tumors. After 51 months median follow-up, median progression-free (PFS) was 8.1 and overall survival 20.2 months, independently of KRAS/NRAS/BRAF or PI3K-pathway mutation status. Patients with TP53 mutations (n = 34; 55%), as well as those with left colon primary tumors (n = 66; 85%), had significantly better PFS, also confirmed in multivariate analysis. Although the clinical trial met its primary endpoint, according to the current standards, the efficacy and tolerability of the drug combination are considered insufficient. Extended genotyping yielded interesting results regarding the significance of TP53 mutations. ClinicalTrials.gov identifier: NCT01215539, Registration date: Sep 29, 2010) [53]. The efficacy of oxaliplatin-based chemotherapy combined with anti-epidermal growth factor receptor (EGFR) monoclonal antibody (mAb) remains controversial in mCRC. A meta-analysis was performed to estimate the effect of adding panitumumab or cetuximab to oxaliplatin-based chemotherapy in RAS wt mCRC patients for the first-line treatment. The primary tumor location is also considered into this meta-analysis. RCT studies were identified by a search of MEDLINE, EMBASE, Cochrane library to October 2017, supplemented by manually retrieving ASCO, ESMO conference abstracts. The pooled hazard ratio (HR) for progression-free survival (PFS) and overall survival (OS), and pooled odds ratios (OR) for the overall response rate (ORR) were calculated by Review Manager 5.3 The results indicated that the addition of anti-EGFR mAbs to FOLFOX regimen in RAS wt mCRC patients for the first-line treatment resulted in considerable improvements in PFS and ORR compared with chemotherapy alone. However, in RAS/BRAF wt patients, no significant differences were observed when anti-EGFR mAb

was added to FLOX or XELOX regimen compared with chemotherapy alone with regard to OS and PFS, whereas FOLFOX+anti-EGFR mAb showed a marked superior OS and PFS. A meta-analysis including TAILOR and PRIME study suggests that primary tumor location (PTL) predicted a survival benefit when adding the EGFR antibody to FOLFOX regimen in RAS wt mCRC patients. However, the HR for PFS and ORR still suggests a benefit from the addition of anti-EGFR mAb in right-sided mCRC patients. So these results suggest anti-EGFR mAb and oxaliplatin are good partners in the FOLFOX regimen. The addition of EGFR antibody to FOLFOX markedly improved efficacy in RAS wt patients with left-sided mCRC. In RAS/BRAF-wt patients, the efficacy is similar. For patients with right-sided tumor, a benefit showing a tendency in favor of anti-EGFR mAb can still seen. The molecular characteristics behind the tumor location need to be more explored urgently [54]. Another study was conducted in Canada where all patients diagnosed before December 2012 and treated with an EGFR antibody for mCRC were identified from the Ontario drug database and linked to the Ontario Cancer Registry and other administrative databases to ascertain baseline characteristics, health services utilization, and outcomes. Multivariable Cox and logistic models were constructed to compare the time to treatment discontinuation (TTD), OS, emergency department (ED) or hospital visits between Cetuximab plus irinotecan versus panitumumab (Camb +I Pamb) Observable confounders were adjusted for using propensity score methods. One thousand and eighty-one patients were identified. Patients, receiving CMab + I, were younger, and had a longer duration of prior irinotecan treatment. The use of CMab + I as compared to Pmab alone was associated with a prolonged time to deterioration (TTD) and an improved OS. The findings for patients over and below the age of 65 were similar [54]. The randomised MACBETH and VOLFI studies showed that a modified FOLFOXIRI regimen in combination with cetuximab or panitumumab, respectively, achieved high therapeutic activity in RAS and BRAF wt patients with an acceptable toxicity profile. Drawing from these evidences, some authors designed TRIPLETE study aiming at comparing two different chemotherapy backbones (mFOLFOXIRI or mFOLFOX6) in combination with panitumumab in the first-line treatment of patients with

RAS and BRAF wt mCRC This was a prospective, open-label, multicentre phase III trial in which initially unresectable and previously untreated RAS and BRAF wt mCRC patients were randomised to receive a standard treatment with mFOLFOX6 plus panitumumab or an experimental regimen with modified FOLFOXIRI plus panitumumab up to 12 cycles, followed by panitumumab plus 5-fluorouracil and L-leucovorin until disease progression. The primary endpoint was overall response rate according to RECIST 1.1 criteria. The relative benefit of chemotherapy intensification when using an anti-EGFR-based regimen in molecularly selected patients is unknown; TRIPLETE study aims at filling this gap of knowledge. The study was sponsored by the Gruppo Oncologico Nord Ovest Cooperative Group and is currently ongoing at 42 Italian centers. (CLINICAL TRIAL INFORMATION: NCT03231722) [55]. The coexpression of pIGF-1R and MMP-7 (double-positive phenotype, DP) correlated with poor overall survival (OS) in KRAS wt (exon 2) mCRC patients treated with irinotecan-cetuximab in second/third line. The authors analyzed two prospective biomarker design trials of newly diagnosed RAS-wt mCRC patients treated with panitumumab-FOLFOX6 (PULSE trial; NCT01288339) or cetuximab plus either FOLFOX6/FOLFIRI (POSIBA trial; NCT01276379). The main exposure was DP phenotype (DP/non-DP), as assessed by two independent pathologists. DP cases were defined by immunohistochemistry as >70% expression of moderate or strong intensity for both MMP-7 and pIGF-1R. Primary endpoint: progression-free survival (PFS); secondary endpoints: OS and response rate. PFS and OS were adjusted by baseline characteristics using multivariate Cox models. They analyzed 67 patients (30 non-DP, 37 DP) in the PULSE trial and 181 patients in the POSIBA trial (158 non-DP, 23 DP). Response rates and PFS were similar between groups in both studies. DP was associated with prolonged OS in PULSE and with shorter OS in POSIBA. A differential effect of anti-EGFRs on survival by DP phenotype was observed. Panitumumab might be more beneficial for RAS-WT mCRC patients with DP phenotype, whereas cetuximab might improve OS in non-DP [56]. Studies investigating combinations of anti-epidermal growth factor receptor monoclonal antibodies such as panitumumab or cetuximab with standard chemoradiation therapy protocols in rectal cancer have yielded disappointing results. Because of the supposed negative interaction of epidermal growth factor receptor inhibition and chemoradiation therapy, we conducted a phase 2 study using single-agent panitumumab in combination with radiation therapy in patients with RAS wild-type locally advanced rectal cancer. Patients with RAS wild-type locally advanced (clinical stage II or III) rectal cancer localized 0 to 12 cm from the anus were eligible for study participation. The primary objective of the study was to determine pathologic complete response (pCR). Secondary objectives comprised assessing the safety, surgical morbidity, clinical response, tumor downstaging, and tumor regression grading according to Dworak. A total of 54 patients with a median age of 58 years were treated. In 3.7% of patients, pCR was achieved. Downstaging of the primary tumor or lymph nodes was seen in 65% of patients. Panitumumab in combination with radiation therapy as neoadjuvant treatment for locally advanced rectal cancer showed a favorable toxicity profile but failed to meet the predefined pCR rate to justify further clinical trials [57].

Gefitinib (ZD1839, Iressa) is a small quinazoline derivative, reversible inhibitor of EGFR tyrosine kinase activity. The antitumor effect is based on its capability to bind strongly to EGFR without inducing EGFR internalization or degradation and without reducing EGFR protein levels. After an accelerated approval process, gefitinib has been approved by the FDA in May 2004 for use as third-line treatment of non-small cell lung carcinoma (NSCLC) refractory to platinum-based and docetaxel-based chemotherapy regimens, but it has been withheld from the USA market since June 2005. Due to the lack of a survival benefit in the ISEL (Iressa survival evaluation in lung cancer) study [58], the FDA restricted the use of gefitinib to patients participating in a clinical trial or continuing to benefit from treatment already initiated [59, 60]. Gefitinib has never been approved in the European Union, but is currently on the market in Japan, Korea, China and several other Asian countries. It is currently an investigational drug in the USA and in the European Union. Different studies have found that three defined regions (exons 18–21) in the EGFR gene are commonly mutated and predict sensitivity to EGFR-TKI. According to Eberhard and others [61, 62], the most common of these mutations in the EGFR gene involve the point mutation, Leucin (L) 858 Arginine (R), in exon21 and an exon 19 deletion, del746-750. According to Sakurada et al. (2006) [63], these comprise up to 86% of EGFR mutations in NSCLCs overall. Concerning the sensitivity of these above-mentioned mutations, it has been suggested that in-frame deletions of EGFR predict higher sensitivity to TKI compared with point mutations in exons 18 and 21 [64, 65]. Specifically, Mitsudomi et al. [64] reported that lung adenocarcinomas harboring the EGFR exon 1 deletion (del746-750) may be more susceptible to gefitinib as compared with tumors with the exon 21point mutation (L858R). Additional mutations in the EGFR gene have been identified, in exons 20 and 18. Since EGFR mutations in NSCLC have a substantially higher frequency in East Asians than in other ethnicities, a relationship between EGFR mutation frequency and ethnicity has been established [66]. These mutations, however, are generally related to an increased sensitivity to EGFR inhibitors, which might not be the case in CRC. There are only a few reports of mutations in the EGFR gene in colorectal tumors and all of these used sequencing after gene amplification. The first report focused on only 31 cases from Milan (Italy) and is the only European study [67]. Of these 31 cases, only one showed a mutation. The second study focused on 33 cases and was performed on Japanese patients from Beppu [68]. Of these 33 cases, ten were detected as having mutations in the EGFR gene. The third focused on 101 Chinese (Shangaï) patients and the authors described only 2 mutations [69]. The last study was conducted by Barber et al. (2004) [70] in Baltimore USA, where only one mutation in293 patients was characterized. Another study on 236 clinical tumor samples from European patients with advanced CRC founds somatic missense mutations in exons 18 and 20 at a frequency of 2.1% and 0.4% respectively. Somatic Single-nucleotide polymorphisms (SNPs) were also found in exons 20 and 21 at a frequency of about 3.1% and 0.4% respectively. In addition, the authors observed the mutated KRAS gene in two cases (2 out of 17 cases, 12%), when EGFR showed a SNP in codon 21 [71]. Moreover, the comparison of the mutational status and clinical parameters would have been of great interest in a larger cohort of patients.

Erlotinib (Tarceva™, OSI-774), a potent epidermal growth factor receptor tyrosine kinase inhibitor (EGFR), and it has been approved by several regulatory agencies worldwide, including the FDA and the European Medicines Evaluation Agency (EMEA), as monotherapy for the treatment of locally advanced or metastatic NSCLC refractory to platinum-based chemotherapy [72]. Furthermore, the erlotinib combined with FOLFOX and bevacizumab resulted in high incidences of toxicities [73]. However, a dose-finding study of erlotinib, combined with capecitabine and irinotecan, identified an acceptable safety profile [74]. Finally, different reports showed that the association erlotinib with chemotherapy and bevacizumab had strong antitumor activity in patients with metastatic colorectal cancer [75-77]. On contrary, another study observed that this combination may be unsuitable for these patients [78]. In a preclinical and phase II study evaluated the efficacy and safety of the combination of cetuximab and erlotinib in mCRC. The activity and mechanism of action of the combination of cetuximab plus erlotinib were investigated in vitro in colorectal cancer cell lines. In the clinical study, patients with chemotherapy-refractory mCRC were treated with cetuximab 400 mg/m^2 as a loading dose and then weekly cetuximab 250 mg/m^2 with erlotinib 100 mg orally daily. The primary end point was response rate (RR), which was evaluated separately in KRAS wt versus KRAS mutant tumors. Secondary end points included toxicity, progression-free survival (PFS), and overall survival. Target accrual was 50 patients, with a one-stage design. Preclinical studies demonstrated synergistic activity of cetuximab and erlotinib cotreatment on growth inhibition of colon cancer cell lines both as a result of enhanced inhibition of the epidermal growth factor receptor pathway and differential effects on STAT3. In the clinical study, 50 patients were enrolled, with 48 patients evaluable for response. The overall RR was 31% with a median PFS of 4.6 months. RR was 41% in KRAS wt tumors, with a median PFS of 5.6 months. There was no response in 11 patients with KRAS mutations. The combination of cetuximab and erlotinib synergistically inhibits growth of colon cancer cell lines, achieves promising efficacy in patients with KRAS wt mCRC, and merits evaluation in further randomized studies [79]. In another study, 82 patients received capecitabine alone (Arm 1) or capecitabine with erlotinib (Arm 2). Median time-to-progression (TTP) in Arm 1 was 7.9 months versus 9.2 in Arm 2. In KRAS-wild type wt patients

TTP was 8.4 and 11.7 months in Arms 1 and 2, respectively. In KRAS-mutated patients TTP was 7.4 and 1.9 months in Arms 1 and 2, respectively. Arm 2 KRAS wt patients, left-sided primaries, had an overall survival of 16.0 versus 12.1 months in right-sided primaries. Adding erlotinib to capecitabine increased TTP by 3.2 months in KRAS wt patients. This study suggests that erlotinib harms patients with KRAS-mutated advanced CRC while it may provide benefit to those with KRAS wt CRC. Further study of EGFR-tyrosine kinase inhibitors in patients with left-sided KRAS-WT CRC is warranted [80]. Additionally, a phase II trial (planned n=58) using second-line therapy for metastatic colorectal cancer with either oxaliplatin-based (mFOLFOX6) or irinotecan-based (FOLFIRI) combination chemotherapy and 100 mg erlotinib daily on days 3-8 after each infusion (days 1 and 2) every 14 days. The FOLFIRI/erlotinib arm met the pre-specified response rate criteria of at least 10% to expand accrual to the intended sample size. The response rate was 18%. In second-line treatment for metastatic colorectal cancer, mFOLFOX6 or FOLFIRI with erlotinib in a sequence-dependent fashion is not feasible despite potential promising activity [81].

Despite advances in the treatment of m CRC, the 5-year survival of patients with this disease remains low. A small proportion of CRCs overexpress the HER2 oncogene and in addition, some studies showed that RAS and BRAF wt mCRC presented a HER2 amplification. This knowledge, in addition to preclinical data providing a rationale for dual anti-HER2 targeted therapy has led to several clinical trials. Recently, several articles reported promising results anti-HER2 therapy in mCRC. Namely, two studies HERACLES-A and MyPathway have shown benefit in a small number of patients with the use of combination trastuzumab-lapatinib and trastuzumab-pertuzumab, respectively [82]. However, the impact of HER2 amplification on prognosis in CRC is unclear. A study by Conradi et al. evalutated HER2 status and prognosis in patients with advanced rectal cancer (cUICC II and III). A total of 264 patients, received standardized multidisciplinary treatment according to protocols of phase II/III trials of the German Rectal Cancer Study Group. HER-2 status was determined in pretherapeutic biopsies and resection specimens using immunohistochemistry scoring and detection of silver in situ hybridization amplification. Tumors with an immunohistochemistry score of 3 or silver in situ hybridization ratios of ≥2.0 were classified HER-2 positive; these results were correlated with clinicopathologic parameters [eg, resection (R) status, nodal status ((*y*)*pN*)], disease-free (DFS) and cancer-specific survival (CSS). Positive HER-2 status was found in 12.4% of biopsies and in 26.7% of resected specimens. With a median follow-up of 46.5 months, patients with HER-2 positivity showed in trend a better DFS and a benefit in CSS. The 5-year survival rate was 96.0% (HER-2 positive) versus 80.0% (HER-2 negative). In univariate and multivariate analyses, HER-2 was an independent predictor for CSS along with the (y)pN status and R status. HER-2 amplification is detectable in a relevant proportion (26.7%) of rectal cancer patients [83]. Conversely, in a study by Park *et* al. HER 2/neu overexpression and gene amplification were examined via semiquantitative standardized immunohistochemical staining and fluorescence in situ hybridization (FISH) in 137 colorectal cancer patients who underwent curative surgery at the Kangbuk Samsung Hospital. Sixty-five out of 137 patients were determined by immunohistochemistry to have overexpressed HER-2/neu protein. HER-2/neu gene amplification was detected in two patients by FISH. Tumors with HER-2/neu overexpression showed higher postoperative recurrence rate. Tumors with HER-2/neu overexpression were associated with poor 3-year and 5-year survival rates. Advanced TNM stage, postoperative recurrence, and overexpression of HER-2/neu were found to be independently related to survival by multivariate analysis. In conclusion, HER-2/neu overexpression may constitute an independent prognostic factor in colorectal cancer patients, and patients exhibiting HER-2/neu overexpression might constitute potential candidates for a new adjuvant therapy which involves the use of humanized monoclonal antibodies [84]. Valtorta et al. developed criteria for ERBB2-positivity (HER2) in colorectal cancer to ensure accurate identification of ERBB2-amplified metastatic colorectal cancer patients. The authors found 100% of concordance between silver in situ hybridization and immunohistochemistry (IHC) using a particular antibody called VENTANA 4B5 antibody. For this reason, they proposed specific criteria for defining ERBB2-positivity in colorectal cancer (HERACLES Diagnostic Criteria). In a phase II trial of trastuzumab and lapatinib in a cetuximab-resistant population, HERACLES Diagnostic Criteria shaped the selection of patients and defined ERBB2 as a predictive marker for response to ERBB2-targeted therapy in metastatic colorectal cancer [85]. In another study, Raghav et al. considered the association of HER2 amplification with RAS and BRAF wt mCRC and they compared progression free survival (PFS) of HER2 amplified mCRC patient with HER2 non amplified patients treated with anti–EGFR-based therapy. They found that 14% of patients with RAS and BRAF wt mCRC had HER2 amplification. In this cohort, median OS and PFS on first line therapy without an anti-EGFR ab (PFS1) was similar between HER-amp and HER2NA (no amplified) patients. A total of 66 patients in cohort 1 received anti-EGFR ab after first line therapy. Median PFS on anti-EGFR ab therapy (PFS2) was significantly shorter in patients with HER amp compared to HER2NA tumors (2.9 v 8.1 m, hazard ratio. These findings were confirmed in cohort 2, in which 69 patients received anti-EGFR ab after first line therapy and median PFS2 was significantly shorter for HER amp patients compared to HER2NA patients with a similar OS and PFS1. In summary, HER2 amplification in mCRC is a predictive biomarker for lack of efficacy of anti-EGFR ab therapy [86]. More recent research found the same data [87]. Richman et al. [88] investigated HER2-amplification/overexpression in stage II-III and IV CRC patients, and the relationships to KRAS/BRAF and outcome. Pathological material was obtained from 1914 patients in the QUASAR stage II-III trial and 1342 patients in stage IV trials (FOCUS and PICCOLO). Tissue microarrays were created for HER2 immunohistochemistry. HER2-amplification was assessed using FISH and copy number variation. KRAS/BRAF mutation status was assessed by pyrosequencing. Progression-free survival (PFS) and overall survival (OS) data were obtained for FOCUS/PICCOLO and recurrence and mortality for QUASAR; 29/1342 (2.2%) stage IV and 25/1914 (1.3%) stage II-III tumours showed HER2 protein overexpression. Of the HER2-overexpressing cases, 27/28 (96.4%) stage IV tumours and 20/24 (83.3%) stage II-III tumours demonstrated HER2 amplification by FISH; 41/47 (87.2%) also showed copy number gains. HER2-overexpression was associated with KRAS/BRAF wt status at all stages: in 5.2% wt versus 1.0% mutated tumours ($p < 0.0001$) in stage IV and 2.1% versus 0.2% in stage II-III tumours ($p = 0.01$), respectively. HER2 was not associated with OS or PFS. At stage II-III, there was no significant correlation between HER2 overexpression and 5FU/FA response. A higher proportion of HER2-overexpressing cases experienced recurrence, but the difference was not significant. HER2-amplification/overexpression is identifiable by immunohistochemistry, occurring infrequently in stage II-III CRC, rising in stage IV and further in KRAS/BRAF WT tumours. The value of HER2-targeted therapy in patients with HER2-amplified CRC must be tested in a clinical [88]. Tu et al. evaluated the protein expression level of human epidermal growth factor receptor 2 (HER-2) using immunohistochemistry (IHC), and assess the association with clinicopathological parameters and the prognosis of patients with CRC. Overexpression of HER-2 and gene amplification were examined with semiquantitative standardized IHC in 878 formalin-fixed paraffin-embedded CRC samples, while 102 of these cases were analyzed with FISH. A total of 102 cases out of the 878 cases, were determined by IHC to overexpress HER-2. Of these, 25 cases were strongly positive (IHC3+), while 77 cases revealed moderate staining (IHC2+). HER-2 overexpression was more frequent in early-stage cases compared with advanced-stage cases of CRC. However, there was no association observed between HER-2 overexpression and clinicopathological parameters. FISH analysis revealed that 64% of the IHC3+ cases had HER-2 gene amplification. By contrast, only 6.5% of the IHC2+ cases, and none of the 20 randomly selected IHC0 or 1+ cases, demonstrated HER-2 gene amplification. Furthermore, no associations were observed between HER-2 overexpression or gene amplification with the survival time. Thus, the present study observed that HER-2 overexpression does not correlate with other clinicopathological data or the survival rate, with the exception of clinical stages. However, IHC3+ and 2+ cases should be further analyzed by FISH to assess the status of the HER-2 gene in CRC. Patients with HER-2 gene amplification may constitute as potential candidates for targeted therapy with trastuzumab [89]. Currently, different antibodies are approved to target the HER2 pathway such as trastuzumab (monoclonal antibody that binds the extracellular domain of HER2), pertuzumab (a recombinant humanized monoclonal antibody which targets the extracellular HER2 dimerization domain and inhibits heterodimerization of HER2 with other HER2 receptors) and lapatinib (dual tyrosine kinase inhibitor targeting the epidermal growth factor receptor (EGFR) and HER-2). Many studies investigated the therapy with the trastuzumab alone or in combination with irinotecan- and oxaliplatin-based chemotherapy in previously treated patients with mCRC and they showed a high antitumor activity [90, 91]. In a phase I/II trial, Rubinson et al. [92] studied the safety and efficacy of pertuzumab and cetuximab in patients with cetuximab-resistant KRAS wt mCRC. Six of the thirteen patients (46%) experienced DLTs, therefore the study was terminated early. Combination pertuzumab and cetuximab in refractory CRC was associated with potential antitumor activity; however, the combination was not tolerable due to overlapping toxicities. An open-label, nonrandomized phase II study included patients with metastatic or locally advanced adenocarcinoma of the colon or rectum with progression by RECIST on or within six months of receiving a fluoridopyrimidine-oxaliplatin- or irinotecan-containing regimen. The study protocol provided lapatinib 1,250 mg orally daily and capecitabine 2,000 mg/m^2 by mouth split into twice-daily dosing for 14 days of a 21 days cycle. Twentynine patients were enrolled in this study. There were no complete or partial responses. 41.4% of patients achieved stable disease as a best response. Median overall survival was 6.8 months, with a 1-year survival rate of 22%, and median progression-free survival was 2.1 months [93]. The seminal HERACLES (HER2 Amplification for Colorectal Cancer Enhanced Stratification) phase 2 trial was conducted by Italian researchers to the efficacy of trastuzumab and lapatinib therapy in patients with treatment-refractory, KRAS codon 12/13 wt and HER2-positive mCRC [94]. This trial was a proof-of-concept, multi-center open-label, phase 2 trial done at four Italian academic cancer centers. They enrolled adult patients with KRAS exon 2 (codons 12 and 13) wt and HER2-positive metastatic colorectal cancer refractory to standard of care (including cetuximab or panitumumab), an Eastern Cooperative Oncology Group performance status of 0 or 1, and at least one measurable lesion. The authors defined HER2 positivity in tumour samples by use of immunohistochemistry and fluorescence in-situ hybridisation in accordance with our previously validated colorectal cancer-specific diagnostic criteria. Eligible patients received intravenous trastuzumab at 4 mg/kg loading dose followed by 2 mg/kg once per week, and oral lapatinib at 1000 mg per day until evidence of disease progression. The primary endpoint was the proportion of patients achieving an objective response (defined as complete response or partial response), which was assessed by independent central review in the intention-to-treat population. This trial is registered with EudraCT, number 2012-002128-33. Between Aug 27, 2012, and May 15, 2015, the authors screened 914 patients with KRAS exon 2 (codons 12 and 13) wt mCRC and identified 48 (5%) patients with HER2-positive tumours, although two died before enrolment. Of these patients, 27 were eligible for the trial. All were evaluable for response. At the time of data cutoff on Oct 15, 2015, with a median follow-up of 94 weeks of 27 patients had achieved an objective response, with one patient achieving a complete response, and seven achieving partial responses; 12 patients had stable disease. In summary, the combination of trastuzumab and lapatinib is active and well tolerated in treatment-refractory patients with HER2-positive metastatic colorectal cancer. Another trial called the HERACLES-RESCUE clinical study is currently investigating the activity of trastuzumab-emtansine (T-DM1), an antibody-drug conjugate consisting of trastuzumab linked to the cytotoxic agent emtansine, in patients with HER2-positive mCRC progressing after trastuzumab plus lapatinib [95]. Another relevant study, the HERACLES cohort B trial is evaluating the clinical activity of lapatinib or pertuzumab in combination with T-DM1 in patients who are HER2–therapy-naïve and have HER2-positive mCRC [96]. Another trial, the MyPathway phase II trial, is investigating the efficacy and safety of pertuzumab in combination with trastuzumab in patients with treatment-refractory mCRC and such patients had an overexpression or amplification of HER2 by gene sequencing and/or by FISH or IHC [97]. The first results showed that twelve patients have achieved partial response, and three have achieved stable disease for longer than four months. The median duration of response is 11.1 mo. Another interesting phase II trial (NCT03457896) evaluated the efficacy of pan-HER inhibitor neratinib plus trastuzumab or neratinib plus cetuximab in patients with quadruple wild-type

(KRAS/NRAS/BRAF/PIK3CA wild-type) HER2-amplified, HER2-nonamplified (wildtype), or HER2-mutated mCRC. The MOUNTAINEER trial will treat patients (with HER2 positive, anti-HER2 targeting therapy-naïve, and RAS wild-type mCRC who have been previously treated with chemotherapy and an antiangiogenic drug) with combination of tucatinib plus trastuzumab in [98]. Tucatinib is an orally bioavailable and HER2 selective small molecule tyrosine kinase inhibitor, In vitro assays were performed to evaluate the combination of tucatinib and trastuzumab in HER2-amplified cell lines by measuring changes in signal transduction (pHER2, pHER3, pAKT) and cell survival. The in vivo activity of tucatinib (50 mg/kg BID) and trastuzumab (20 mg/kg Q3D) was evaluated alone, or in combination, in established HER2-amplified tumor models, including PDX models of colorectal, esophageal and gastric cancers. As a single agent, or in combination with trastuzumab, tucatinib demonstrated significant anti-tumor activity, including tumor regressions, in the N87 gastric cancer cell line xenograft model and in PDX models of HER2 amplified colorectal, esophageal and gastric cancers. The combination of tucatinib and trastuzumab was consistently more active than either single agent alone and resulted in tumor growth inhibition from 85-139%, including complete tumor regressions in HER2+ gastric PDX models. In conclusion, the activity of tucatinib in HER2-amplified colorectal, esophageal and gastric tumor xenograft models supports the exploration of using tucatinib to treat HER2+ gastrointestinal cancers in the clinical setting [99]. Furthermore, another trial conducted by The Southwestern Oncology Group is evaluating the efficacy of trastuzumab in combination with pertuzumab versus cetuximab and irinotecan in patients with HER2 amplified, RAS and BRAF wild-type metastatic or unresectable colorectal cancer who have progressed on at least one prior regimen of systemic chemotherapy (S1613, NCT03365882). At the end, ZW5 is an engineered bi-specific monoclonal antibody that targets two different epitopes of the human tumor-associated antigen (TAA) epidermal growth factor receptor 2 (HER2) with potential immunomodulating and antineoplastic activities and it may be effective and well tolerated in patients with a variety of HER2-positive cancers. In a phase I trial, patients treated with the drug—most of whom had gastroesophageal or colorectal cancer—had an objective response rate of 41% and mild side effects [100].

Afatinib [Giotrif] is an orally administered, irreversible inhibitor of the ErbB family of tyrosine kinases. A randomised phase II trial was conducted in patients with KRAS wt mCRC with progression following oxaliplatin- and irinotecan-based regimens using afatinib and cetuximab Patients with KRAS wt tumours were randomised 2:1 to afatinib (40 mg/day, increasing to 50 mg/day if minimal toxicity) or cetuximab weekly (400 mg/m2 loading dose, then 250 mg/m2/week) according to number of previous chemotherapy lines. All patients with KRAS-mutated tumours received afatinib. Primary endpoints were objective response (OR) for the wild-type group and disease control for the KRAS-mutated group. Secondary endpoints were progression-free survival (PFS) and overall survival (OS). Patients with KRAS wt tumours received afatinib or cetuximab. Unconfirmed and confirmed ORs were 3% and 0% for afatinib versus 20% and 13% for cetuximab. Median PFS was 46.0 and 144.5 days for afatinib and cetuximab, respectively. Median OS was 355 days with afatinib but not reached for cetuximab. In the KRAS-mutated group, five patients achieved confirmed disease control; no ORs were reported. Median PFS and OS were 41.0 and 173 days, respectively. The efficacy of afatinib was inferior to cetuximab in patients with KRAS wt mCRC. In patients with KRAS-mutated tumours, disease control was modest with afatinib. Afatinib had a manageable safety profile [101]. In another trial, the authors investigated dual targeting with cetuximab and afatinib in chemotherapy refractory KRAS wt mCRC. This study was conducted in KRAS wt mCRC patients who failed chemotherapy with irinotecan and oxaliplatin. Due to low accrual and the frequent use of an anti-EGFR antibody in the first- or second- line, the study was amended, after 19 inclusions, to allow pre-treatment with anti-EGFRs. This multicenter, prospective, open phase II trial randomized patients (2:1) to cetuximab (500mg/m^2/two weeks) and oral afatinib (40mg/day) (Arm A) or cetuximab alone (500mg/m^2/two weeks) (Arm B). Treatment continued until progression (evaluated every six weeks using RECIST 1.1) or limiting toxicity. At progression (Arm B) afatinib could be added to cetuximab. The primary objective was the non-progression rate at six months. The authors randomized 51 patients in Arm A and 24 in Arm B. The non-progression rate at six months was 17.0% in Arm A and 20.8% in Arm B. Median PFS (CI95%) and OS were 4.1 months and 13.6 months in Arm A; and 2.7 months and 13.4 months in Arm B. The objective response was 26.0% in Arm A and 8.3% in Arm B. The median PFS has to be interpreted considering the criterion allowing pre-treatment with anti-EGFRs. Cross-over in Arm A could explain the similar median OS in both arms. Additional data will be presented (NRAS, subgroup of pts not pre-treated with anti-EGFRs, pts treated with cross-over). Clinical trial information: NCT01919879 [102]. Finally, a case report was presented: a KRAS/NRAS/BRAF mCRC wt patient who has been previously treated with FOLFIRI (fluorouracil, leucovorin, and irinotecan), XELOX (capecitabine and oxaliplatin), cetuximab and bevacizumab, and then received the next generation sequencing (NGS) and whose metastatic subcutaneous nodule was resected to generate patient-derived xenograft (PDX) models. The NGS revealed HER-2 amplification as well as an activating mutation S310F and PDX models tested several drugs finding that afatinib was the optimal agent with notable efficacy and well tolerance among 6 regimens. Therefore, this patient started to take afatinib orally and achieved 3 months progression-free survival (PFS) and relief of clinical symptoms without severe adverse effects. In conclusion, NGS and PDX models have great significance for precision and individualized medicine in the mCRC treatment, especially for patients whose diseases have been progressed after multiline standard therapies [103].

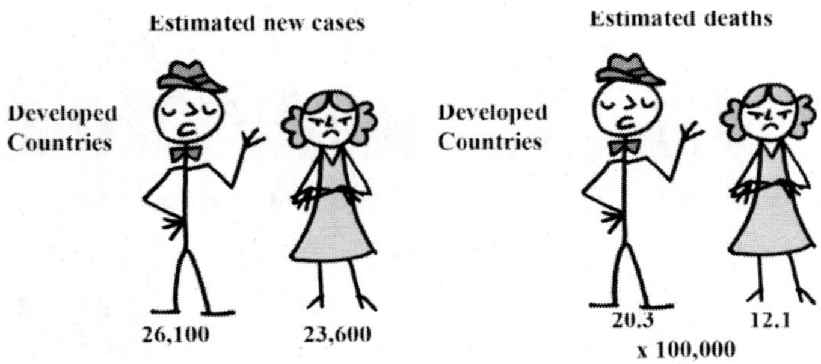

Figure 1. Estimated new cases and deaths in developed countries [2, 3].

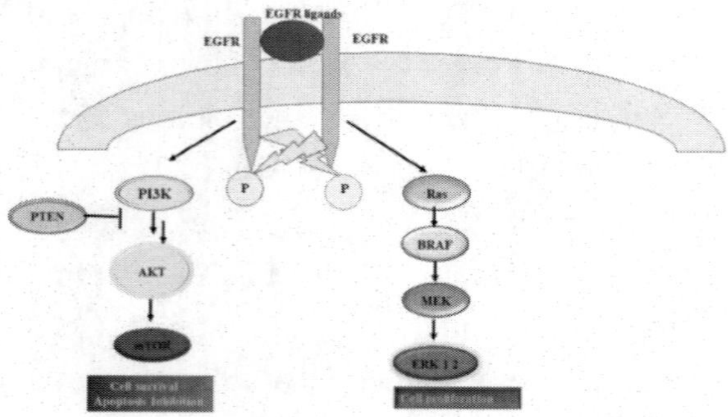

Figure 2. The EGFR pathway.

PRIMARY RESISTANCE TO EGFR THERAPY

Retrospective analyses found that patients with CRC carrying activating KRAS gene mutations did not respond to cetuximab therapy [104]. On the contrary, other studies reported that cetuximab or panitumumab had an antitumoral activity in patients with KRAS mutated tumors [105, 106]. Specifically, patients with KRAS mutations in codon 12 or 13 did not benefit from anti-EGFR therapy [107]. while patients with KRAS exon 2 mutations resulted resistance to the anti-EGFR drugs cetuximab and panitumumab [108]. However, the data regarding the codon 13 mutation (G13D) of KRAS gene were contrasting [106, 109]. The last study reported the analysis on 579 patients with chemotherapy-refractory CRC were treated with cetuximab from different studies. Such study observed a positive association between KRAS G13D mutations and OS benefit with cetuximab treatment. Finally, another analysis that pooled data sets from the CRYSTAL and OPUS studies showed a similar result [109]. On contrary, a retrospective analysis reported that cetuximab treatmen in 110 patients with KRAS G13D mutations did not have anti-cancer activity [110]. In according with these data, a pooled analysis of three randomized phase III trials showed that patients with KRAS G13D mutations did not benefit from panitumumab therapy [111]. An updated randomized phase III CRYSTAL study considered patients with KRAS exon 2 wt and additional RAS mutations *(KRAS* exons 3 and 4;

NRAS exons 2, 3 and 4). In this case, the patients did not respond to the additional cetuximab treatment, while patients with no RAS mutations showed a significant PFS and OS when cetuximab was added to chemotherapy. Furthermore, a recent subset analysis of data from the PRIME trial showed a consistent result that patients with mutations close to the KRAS exon 2 could not have some benefit during first-line chemotherapy in association with panitumumab [112]. In this study, the authors observed that the patients with tumors bearing mutations in KRAS exon 3 (at codon 61) or 4 (at codons 117 and 146), or in NRAS exon 2 (at codons 12 and 13), 3 (at codon 61), or 4 (at codons 117 and 146) had a worse PFS and OS when treated with FOLOFX4 in combination with panitumumab. The mCRC Patients with BRAFmutations did not respond to cetuximab therapy in with mCRC [105, 113]. A similar result was demonstrated in a more recent multicenter randomized PICCOLO trial [114]. On the contrary, the first line randomized studies (OPUS and Cetuximab combined with Irinotecan-CRYSTAL) showed a trend in favor to cetuximab treatment for BRAFV600E patients. Finally, in another study, the patients with refractory BRAFV600-mutant mCRC were treated with a selective RAF kinase inhibitor (encorafenib) in combination with cetuximab and with or without a PI3Kα inhibitor (alpelisib). The primary objective was to determine the maximum tolerated dose (MTD) or a recommended phase II dose. Dose-limiting toxicities were reported in 3 patients receiving dual treatment and 2 patients receiving triple treatment. The MTD was not reached for either group and the phase II doses were selected as 200 mg encorafenib (both groups) and 300 mg alpelisib. Combinations of cetuximab and encorafenib showed promising clinical activity and tolerability in patients with BRAF-mutant mCRC. Herein, the authors demonstrate that dual- (encorafenib plus cetuximab) and triple- (encorafenib plus cetuximaband alpelisib) combination treatments are tolerable and provide promising clinical activity in the difficult-to-treat patient population with BRAF-mutant mCRC [115]. These findings therefore, demonstrate a strong causal relationship between the presence of the BRAF V600E mutation and resistance to anti-EGFR therapy in mCRC. Furthermore, the inactivation of the PTEN phosphatase, or the activating mutations of the PIK3CA p110 subunit can deregulate the PI3K/PTEN/AKT pathway. Two studies have reported that lack of response to cetuximab in mCRC patients were attributable to PIK3CA mutations or PTEN loss [116, 117]. Specifically, in the last study, the authors observed that 12 patients with PTEN-positive metastases responded less to cetuximab-based treatment, when compared with one patient with PTEN-negative metastases. Another study reported that PTEN-positive tumors have a better outcome than negative ones, two other ones failed to confirm this observation [118]. Furthermore, PIK3CA mutations in exon 9 and exon 20 play an important role in the therapy of CRC as shown by biochemical studies since exon 9 mutations activate gain of function through RAS-GTP binding, while exon 20 mutations is independent of interaction with RAS-GTP [119]. A large retrospective study investigated the effect of PIK3CA mutations on the response to cetuximab-based therapy [120] and it found that the patients with the KRAS wt and PIK3CA exon 20 mutations had lower response rates when compared with PIK3CA wt patients, while exon 9 mutations showed no significant effect. Furthermore, a meta-analysis from 13 retrospective cohort studies found that PIK3CA exon 20 mutations patients did not respond to anti-EGFR MoAbs [121]. Also, the Janus family of tyrosine kinases (JAK) and the signal transducer and activator of transcription (STAT) family can be involved resistance to anti-EGFR moAbs [122]. Many findings suggested the involvement of SMAD/STAT3 signaling pathway, the NF-kB pathway, the RAS-mitogen- activated protein kinase/Snail/Slug and microRNAs in the development of colorectal cancers via epithelial-to- mesenchymal transition (EMT) [123]. Thus, EMT appears to be involved in the mechanism of anti-EGFR therapy resistance. Buck et al. found a strong association between E-cadherin (epithelial marker) expression and growth inhibition by EGFR inhibitors in CRC cells [124]. In particular, some colorectal tumor cell lines resulted to be insensitive to EGFR inhibition since they had lost or mutated the epithelial junction constituents E-cadherin and gamma-catenin, had lost homotypic adhesion, and often gained proteins associated with an epithelial to mesenchymal-like transition, such as vimentin, zeb1, or snail. In matched pairs of colorectal tumor cells, the epithelial lines showed an average 7-fold greater sensitivity than mesenchymal-like lines. Moreover, another study showed that the antisense mixed backbone oligonucleotides (AS MBOs) directed against TGF-alpha, AR and CR. was able to overcome the resistance of mesenchymal-like CRC cells to anti-EGFR drugs [125]. So far, there is no clinical evidence of the correlation between EMT and resistance to EGFR inhibitors in mCRC. For this reason, it could be important to carry out new studies to better understand mechanism of the EMT induced resistance to EGFR inhibitors.

SECONDARY RESISTANCE TO EGFR THERAPY

The molecular bases of secondary resistance to cetuximab and panitumumab in colorectal cancer are poorly understood. In a report, the author found mutant KRAS DNA in 28 patients treated with panitumumab [126]. However, nine developed detectable mutations in such gene after the treatment. In 2010, Bouchahda et al. found the first tumor KRAS mutations in CRC after the development of resistance to cetuximab [127]. The two KRAS mutations at codon 13 and 12 were detected in the metachronous liver metastatic tissues after cetuximab treatment. A similar study was performed by Misale et al., and the authors found KRAS mutations in six of the 10 patients after cetuximab therapy using the molecular profiles of relapsed tumors from CRC patients [128]. Furthermore, they reported *KRAS* amplification in one the patient receiving anti-EGFR moAbs. Moreover, it was possible to detecte mutations in NRAS and BRAF in association with secondary resistance. For instance, a preclinical model, cetuximab- and/or panitumumab-resistant CRC cell lines developed alterations at the known hotspots of the *KRAS* gene, NRAS and BRAF mutations [128]. Bertotti et al. reported an amplification of HER2 as another possible mechanism of resistance to cetuximab in mCRC with wt KRAS/NRAS/BRAF/PIK3CA genes [129]. Furthermore, Yonesaka et al. identified the amplifications of HER2 and increased expression level of heregulin in cetuximab-resistant cell lines [130]. Another molecule as Insulin-like Growth Factor Receptors (IGFR) is involved in acquired resistance to EGFR targeting therapies: a study showed that the gefitinib-resistant A431 cells presented an increased signaling via the Insulin-like Growth Factor 1 Receptor/Insulin Receptor Substrate 1/(IGF-1R/IRS-1) /PI3K pathway [131]. Different early clinical analyses were conducted in order to demonstrate the anti-tumor activity of IGFR inhibitors alone or the combination with anti-EGFR moAbs in patients with anti-EGFR moAb-refractory mCRC [132, 133]. In particular, one randomized phase II/III study, 344 eligible patients with *KRAS* wt tumors were randomly treated with dalotuzumab (IGF1R inhibitor) or placebo in combination with cetuximab and irinotecan. The data found that the addition of dalotuzumab did not improve PFS or OS compared to the placebo group. However, further analysis on IGF1 expression showed an improvement in PFS and OS in patients with IGF-1 positive tumors with the addition dalotuzumab. Moreover, the results reported shorter PFS and OS in placebo group patients respect to IGF-1 positive tumors group patients. These synergistic anti-tumor effects further support a role for IGF-1R as a mediator of resistance to anti-EGFR agents. EGFR-MET connections play an important role in acquired resistance to cetuximab in CRC cells after activation of the MET pathway induced by the overexpression of TGF-α [133]. In addition, in 2011, Liska et al. showed that an increased level of HGF-induced MET activation could confer cetuximab resistance to CRC cells [134]. In particular, the authors observed that could rescue CRC cells did not undergo apoptosis or cell cycle arrest triggered by cetuximab since HGF-mediated MET activation was able to restore the cellular proliferation through the AKT and MAPK pathways [134]. Furthermore, another study on tumor tissues from patients with acquired cetuximab or panitumumab resistance found tthe *MET* amplification in more than 40% of the cases [135]. Additionally, such amplification was rarely found in pre-treatment tumor tissues. In particular. *MET* amplification could also be involved in primary resistance to anti-EGFR moAbs in CRC. Therefore, the findings of these studies suggested that amplification of *MET* was assocciated with a negative response to cetuximab [136]. However, MET amplification was observed in only 1% of untreated mCRC cases [136]. In 2011, Montagut et al. found an acquired mutation in the extracellular domain of EGFR (S492R) in patients treated with cetuximab in mCRC [137]. The authors observed a disease progression after cetuximab terapy in 10 patients: two patients developed a S492R mutation. However, such mutation was not detected in their pretreatment biopsies. Addionally, the 1476C > A or the 1474A > C mutation in the gene region encoding for the extracellular domain of EGFR caused the substitution of serine to arginine at amino acid 492 (S492R) [137]. Such mutation provoked a decreased affinity betweent the receptor and the ligand and inhibited the binding with cetuximab, while it did not interfere with binding of panitumumab to EGFR. Indeed, one patient with this mutation and resistant to cetuximab- responded very to subsequent treatment with panitumumab. In addition, the tumor samples collected from 156 mCRC patients that did not undergo any therapy did not show the same mutations. In according with this study, Esposito et al. obtained the same result in a larger cohort of patients [138]. Overall, these evidences strongly supported that the S492R mutation is a mechanism of acquired secondary resistance to cetuximab. Finally, many reports suggested that Sphingosine Kinase1/Sphingosine 1-Phosphate (SphK1/S1P) pathway is involved in the resistance of cancer cells [139, 140]. An overexpression of SPHK1 was found in cetuximab resistant colorectal cancer cell lines and the initial sensitivity to drugs was replaced through SphK1 knockdown by both pharmacological inhibitor or silencing via siRNA. Moreover, treatment of resistant cells with fingolimod (FTY720), a S1P receptor (S1PR) antagonist, resulted in re-sensitization to cetuximab in vivo, with inhibition of tumor growth, induction of cancer cells apoptosis, and prolongation of mice survival [141].

Chronic Myeloid Leukemia (CML)

Chronic myeloid leukemia (CML) is a malignancy arising from transformation of the hematopoietic stem cell, which typically evolves through 3 distinct disease stages: an indolent chronic phase (CP), characterized by the accumulation of mature granulocytes and myeloid precursors in the bone marrow and in the peripheral blood; an accelerated phase, characterized by an increase in disease burden and in the frequency of progenitor/precursor cells; and an acute phase called blast crisis (BC), marked by increasing numbers of differentiation-arrested blast cells [142]. The instigating factor in the pathogenesis of chronic myeloid leukemia (CML) is the formation of the Philadelphia chromosome resulting from the reciprocal translocation between chromosomes 9 and 22 (t(9;22)(q34;q11)), which is associated with the de novo creation of the BCR-ABL fusion oncogene [143, 144]. In the ABL gene, the break point is generally located upstream of the second exon (a2), whereas in the BCR gene the breakage takes place usually in one of the three regions called major (M-bcr), minor (m-bcr), and micro-bcr (μ-bcr) break point regions. Depending on the location of the chromosome breakage in the BCR gene, three different types of BCR–ABL proteins, differing in mass as well as biological properties, can be formed .Many studies have demonstrated that the majority of CML patients have a p210 BCR–ABL gene (M-bcr), in which the fusion is located downstream of the 14 or 13 exons of the BCR gene resulting in the creation of mRNA transcripts, which have an e14 and/or an e13 junction, and thus a 210 kDa chimeric protein is produced by third break point in the region of the BCR gene called μ-bcr results in the transcription of an e19/a2 mRNA that codes a 230 kDa BCR–ABL protein. This form of fusion protein is associated with the rare Ph-positive chronic neutrophilic leukemia from this mRNA. The smallest of the fusion proteins, p190 BCR–ABL, is developed as a result of the minor break point region (m-bcr) of the BCR gene leading to the transcript e1a2. The p190 BCR–ABL form is mainly associated with Ph-positive acute lymphoblastic leukemia and rarely appears in patients with CML and might correlate with an aggressive course of the disease, instead the form of p210 BCR–ABL that is found in over 90% of CML patients [145[. Several treatments for CML with tyrosine kinase inhibitors (TKIs) of BCR-ABL1 have revolutionized CML therapy [146].

DEVELOPMENT OF TYROSINE KINASE INHIBITORS (TKIS)

In the early 1990s, Brian Druker collaborated with Novartis expecting to identify such a tyrosine kinase inhibitor that can be orally administrated, and they found that a 2-phenylaminopyrimindin derivative, CGP57-148B, now widely known as imatinib (IM), with a brand name Gleevec, capable of potently inhibiting BCR-ABL and killing CML cells while leaving normal cells unaffected [147, 148]. Imatinib mesylate (Gleevec™, STI-571) was the first TKI, U.S. Food and Drug Administration (FDA) approved as a selective and potent inhibitor of BCR-ABL [147]. Such drug was able to inhibit BCR-ABL kinase activity by occupying the ATP-binding pocket through forming hydrogen bonds and blocking its access to ATP, thereby locking it at inactive confirmation and precluding subsequent substrate phosphorylation and activation of downstream signaling pathways [149]. Imatinib was first used in CML in patients who had developed resistance or intolerance to interferon-α (IFN-α). Among 532 such patients treated with imatinib, a complete cytogenetic response (CCyR) was achieved in 60%. The estimated 5-year survival rate was 76% [150, 151]. At 18 months the rates of Mcg was 87.1% with the imatinib compared to the previous therapy with Mcg of the 34.7%, besides, disease progression accelerated or blast phase CML (AP and BP) was 96.7% with the imatinib compared to the previous therapy by 91,5% and importantly imatinib was better tolerated [152]. At the 5-year follow up 87% of patients had achieved a CCyR, with an estimated OS of 89% [153]. At 8 years of follow-up on the IRIS study, 45% of patients had discontinued treatment due to adverse events/safety (6% unsatisfactory therapeutic outcomes (16%), death (3%) or other reasons (17%), The OS rate was 85%, with the annual rates of progression to AP or BP in years 4 to 8 after imatinib therapy being: 0.9%, 0.5%, 0%, 0%, and 0.4%, respectively. Progression to AP or BC was noted only in 3% of patients who achieved complete cytogenetic response (CCgR) and in none of patients who achieved major molecular response (MMR). While conventional response criteria for CML have followed the optimal achievement of a complete hematological response (CHR) by 3 months, CCgR by 9-12 months and a MMR by 18 months. Although Imatinib shows great efficacy in controlling CML progression in chronic phase patients, it is less effective in patients with advanced disease, or even ineffective in blast crisis patients [154]. Collectively, these unresponsive events were considered to be imatinib resistance [155]. In certain cases, patients don't respond to imatinib at all, which is called primary resistance [156] and is largely due to unknown genetic alterations that shift the dependence away from BCR-ABL for survival. Imatinib is also vulnerable to MDR1, p-Glycoprotein, mediated drug efflux [157]. In other cases, which are regarded as secondary resistance [156] patients may respond to IM at first but lose established response overtime. These cases often involve BCR-ABL mutations that block Imatinib binding. Although the results observed with imatinib were highly effective TKI for ~60% of CML patients [158], in many patients experience some impairment in quality of life, such as fluid retention (periorbital and peripheral), muscle cramps, or gastrointestinal disturbances (nausea, vomiting and diarrhea) [152, 153, 159]. Furthermore there are studies that signal organ toxicity and cardiac dysfunction after prolonged therapy with imatinib [160] but about 25% of patients showed primary and secondary resistance, in fact the hematologic resistance has manifested itself in 2–4% of cases, while cytogenetic resistance is more common, occurring in 15–25% of patients. Recent work suggests that primary resistance may be associated with increased transcript levels of the drug metabolism gene prostaglandin-endoperoxide synthase 1/cyclooxgenase 1 (PTGS1/COX1), and this may serve as a biomarker to distinguish patients with primary resistance to imatinib [161]. Several mechanisms of resistance to imatinib have been described, these can be classified into two categories: BCR-ABL-dependent and BCR-ABL independent. In fact, the first group includes amplification or overexpression of BCR-ABL or its protein product [162] and point mutations of the ABL sequence [163]. Recent studies have found that BCR-ABL-dependent mechanisms are more common, particularly point mutations, which have been identified in approximately 50% of patients who develop clinical resistance to imatinib [164]. Have been described more than 90 different mutations and occur in any of the different relevant domains of the kinase, including the ATP-binding domain (also known as P-loop), the catalytic domain, the activation loop, and amino acids that make direct contact with imatinib. Most of the clinically relevant mutations develop in a few residues in the in the P-loop (G250E, Y253F/H, and E255K/V), the contact site (T315I), and the catalytic domain (M351T and F359V) [165]. Branford S and others have asserted that the P-loop mutations have been suggested to carry an increased risk of rapid blastic transformation and short survival [166], although other authors did not support this study [167]. Many studies have verified that any patients were resistant to standard dose imatinib. In a retrospective analysis of patients enrolled in the IRIS trial. Kantarjian et al. reported that among 106 patients who required dose escalation due to resistance to standard dose therapy, the OS rates were 89% and 84%, respectively, at 3 years from dose escalation [168]. In another study from MDACC (M.D. Anderson Cancer Center (MDACC), 84 patients with CML-CP were dose escalated to imatinib 600–800 mg/day after developing hematologic failure (n = 21), or cytogenetic failure (n = 63) to standard dose imatinib [169]. Many studies have affirmed that several Phase II studies examined the role of a higher dose of imatinib (800mg) upfront in the treatment of patients with CML. The Tyrosine Kinase Inhibitor Optimization and Selectivity (TOPS) study was a phase 3 trial comparing the efficacy and safety of high-dose (800 mg/day) with standard-dose imatinib (400 mg/day) in patients with newly diagnosed CML-CP [170]. The result of the study was rate of major molecular response (MMR) at 12 months of therapy. Subsequently, a 24-month update on the TOPSIt appears that there was no significant difference between the 800 mg/day and 400 mg/day arms for either the CCyR (76% vs 76%, respectively; P = 1.00) or MMR rate (51% vs 54%, respectively; P = .626). Adverse events tended to be more common among patients in the 800mg/day arm vs the 400-mg/day arm, as was the rate of discontinuation due to adverse events (12% vs 5%, respectively). The results from TOPS study were confirmed in a randomized trial Gruppo Italiano Malattie Ematologiche dell'Adulto (GIMEMA) 021/ELN (021/ELN) assessing the efficacy of imatinib 800 mg/day vs 400 mg/day as front-line therapy in high risk Sokal patients [171]. The primary study endpoint of CCyR at one year was not significantly different between patients treated with imatinib 400 mg/day (61%) vs 800 mg/day (64%). There was a trend toward higher rates of MMR with 800 mg/day compared with 400 mg/ day, but the differences were not statistically significant. Furthermore, the compliance was lower in the 800-mg arm (62% received doses >600 mg) compared with the 400-mg arm (87% received doses >350 mg). Although, several studies have shown improved CCyR and MMR with a higher dose of imatinib, but imatinib at a dose of 400mg daily is still the preferred regimen of choice in newly diagnosed patients with CML-CP (chronic phase). Many studies have demonstrated that with time, 20-30% of patients develop TKI resistance, commonly attributable to BCR-ABL1 kinase domain mutations, some patients fail therapy despite inhibition of BCR-ABL1, implicating activation of alternative resistance mechanisms, and the 5-10% discontinue therapy secondary to poor tolerability [172]. This analysis has led to the development of newer TKIs' such as, dasatinib, nilotinib, bosutinib and ponatinib. Crystallography facilitated the synthesis of second-generation TKIs, such as nilotinib (Tasigna; Novartis Pharmaceuticals), dasatinib (Sprycel; Bristol-Myers Squibb) and bosutinib (Bosulif; Pfizer). The second-generation TKIs were found to exhibit selectivity for the BCR-ABL1 mutations not affected by imatinib [173, 174]. However, none of these three TKIs could suppress the BCR-ABL1 T315I mutation that emerges in up to 20% of refractory patients [175]. Although the third- generation TKI ponatinib (Iclusig; ARIAD Pharmaceuticals) inhibits this mutation, additional mutations have emerged and many of them defy inhibition even by this broad-target TKI [176] (Table 1).

Table 1. Summarizes information on TKIs currently available for CML treatment

	Compund	FDA approval	First-line therapy	Alternative name	Brand name	Company
I	imatinib	2001	Yes	CGP571488. STI-571. Imatinib mesylate	Glivec	Novartis Pharmaceuticals
II	dasatanib	2006/2010	Yes	BMS 354825	Sprycel	Bristo-Meyers Squibb
	nilotinib	2007/2010	Yes	AMN107	Tasiga	Novartis Pharmaceuticals
	bosutinib	2012	No	SKI-606	Bosulif	Pfizer
III	ponatinib	2012 (discotinued) 2014	No expect for T315-positive CML	AP24534, ponatinib hydrochloride	Iclusig	Ariad Pharmaceuticals

Dasatinib (Sprycel) Bristol-Myers Squib, Princeton, NJ) is an orally bioavailable, multikinase inhibitor that is be 325-fold more potent than imatinib against unmutated BCR-ABL [177]. Dasatinib, a second-generation BCR-ABL1 tyrosine kinase inhibitor, is approved for the treatment of chronic myeloid leukemia and Philadelphia chromosome-positive acute lymphoblastic leukemia, both as first-line therapy and after imatinib intolerance or resistance. It is a potent second-generation BCR-ABL1 tyrosine kinase inhibitor (TKI) approved at 100 mg once daily (QD) as first-line therapy in patients with chronic myeloid leukemia in chronic phase (CML-CP), and in patients with CML-CP who are resistant to or intolerant of prior therapy. Dasatinib is also a highly potent inhibitor of SRC family kinases including FRG, FYN, HCK, LCK, LYN, and YES, and has activity against additional tyrosine kinases, such as KIT and PDGFR [178]. Dasatinib is also approved at 140 mg QD in patients with accelerated or blast phase CML (CML-AP/BP) or Philadelphia chromosome-positive acute lymphoblastic leukemia (Ph^+ ALL) with resistance to or intolerance of prior therapy [179]. Dasatinib exclusively targets the active confirmation of ABL kinase, but with less stringent conformation requirement and reduced selectivity [179]. Several authors have demonstrated that [DASISION (CA180-056] in the phase 3 dasatinib versus imatinib study in treatment-naive CML patients (DASISION [CA180-056]; clinicaltrials.gov identifier 00481247), were randomized to receive either 100 mg QD dasatinib (n=259) or 400 mg QD imatinib (n=260) with newly diagnosed CML. The primary endpoint was the proportion of patients with confirmed complete cytogenetic responses (CCyR) by 12 months. Patients included in this trial had a minimum of 5 years of follow up [180]. In the phase 3 dose-optimization study in imatinib-resistant or -intolerant CML-CP patients (CA180-034; clinicaltrials.gov identifier 00123474), were randomized 670 patients with CML-CP intolerant of or resistant to imatinib to receive dasatinib 100 mg QD (n=167), 140 mg QD (n=167), 50 mg twice daily (BID; n=168), or 70 mg BID (n=168). (20–22). A subset of patients modified their dose over the course of the study; however, analyses of data were performed according to each patient's initial randomization. The primary endpoint was the percentage of patients with major cytogenetic response (MCyR) after a minimum follow up of 6 months. Patients included in this report had a minimum of 7 years of follow up [181]. Population of patients with Ph+

leukemia. N=2712) were treated with first- or second-line therapy with dasatinib 15 mg to 240 mg QD in 1 of 11 phase 1, 2, or 3 trials were included in these analyses. Study of patients (N=2712) with Ph^+ leukemia who were treated with first-or second-line therapy with dasatinib 15 mg to 240 mg QD in 1 of 11 phase 1, 2, or 3 trials were included in these analyses [182-186]. Several studies have shown as a risk factor the pleural effusion in dasatinib-treated patients in relation advanced age, advanced disease, heart disease, preexisting hypertension, hypercholesterolemia, autoimmune disease, rash-drug dose and schedule, and the presence of lymphocytosis [187-191].

Nilotinib is a tyrosine kinase inhibitor (TKI) approved for the treatment of adult patients with newly diagnosed Philadelphia chromosome–positive (Ph+) chronic myeloid leukemia in chronic phase (CML-CP) or imatinib-resistant or imatinib-intolerant Ph+ CML in CP or accelerated phase (AP) [192-193]. It was developed with rational drug design based on the structure of ABL-IM complex to overcome resistance related to BCR-ABL mutations [194]. Nilotinib binds to the inactive confirmation of ABL kinase domain through lipophilic and weak van der wall interactions and blocks its catalytic activity [194], and it is not dependent on OCT1 or P-glycoprotein for influx and efflux [195]. Nilotinib is 10-30-fold more potent than IM in inhibiting BCR-ABL activity and proliferation of BCR-ABL+ cells, and it is effectively against 32/33 BCR-ABL mutations at physiologically relevant concentration [196]. Notably, a limited number of P-loop associated mutations, such as E255K and Y253H, causes resistance to Nilotinib, but can be suppressed by increased drug concentration [169]. However, the loss of effectiveness against T315I mutation might be a consequence of the steric hindrance between the isoleucine-methyl group of BCR-ABL and 2-methylphenyl phenyl group of Nilotinib. Evaluating Nilotinib in Clinical Trials–Newly Diagnosed Patients (ENESTnd) is a randomized phase 3 study evaluating nilotinib 300 or 400 mg twice daily vs imatinib 400 mg once daily in patients with newly diagnosed CML-CP [193, 198]. Several studies have observed that the rate of major molecular response (MMR; BCR-ABL$\leqslant$0.1% on the International Scale (the rate of major molecular response (MMR; BCR-ABL$\leqslant$0.1% on the International Scale (BCR-ABLIS)) at 12 months, was met, with significantly higher rates of MMR at 12 months in the nilotinib 300-mg (44%) and 400-mg (43%) twice-daily arms than in the imatinib arm (22% P <0.001 vs each nilotinib arm).)) at 12 months, was met, with significantly higher rates of MMR at 12 months in the nilotinib 300-mg (44%) and 400-mg (43%) twice-daily arms than in the imatinib arm (22% P <0.001 vs each nilotinib arm [198]. The adults (N=846) within 6 months of CML-CP diagnosis and without previous CML therapy (except hydroxyurea and/or anagrelide or ⩽2 weeks of imatinib) and an Eastern Cooperative Oncology Group performance status of ⩽2 were randomized (stratified by Sokal risk score at baseline) 1:1:1 to receive nilotinib 300 mg twice daily (n=282), nilotinib 400 mg twice daily (n=281) or imatinib 400 mg once daily (n=283). Molecular responses were assessed by BCR-ABL/ABL transcript ratios using real-time quantitative polymerase chain reaction (RQ-PCR) at a central laboratory (Molecular MD, Portland, OR, USA) standardized to the International Scale (IS), By 5 years, 217 (77.0% 95% CI, 71.6–81.7%), 217 (77.2% 95% CI, 71.9–82.0%) and 171 (60.4% 95% CI, 54.5–66.2%) patients in the nilotinib 300-mg twice-daily, nilotinib 400-mg twice-daily and imatinib arms, respectively, achieved MMR. Furthermore, among patients in each arm who achieved MMR at any time before the data cutoff, 13 of 218 (6.0%), 16 of 220 (7.3%) and 17 of 173 (9.8%), respectively, had a confirmed loss of first MMR (7, 9 and 10 of these patients, respectively, later regained MMR). Very few patients without MMR by 5 years remained on core treatment at the data cutoff (5, 9 and 10 patients in the nilotinib 300-mg twice-daily, nilotinib 400-mg twice-daily and imatinib arms, respectively, including 3, 0 and 2, respectively, with atypical transcripts at baseline). Addionally the frequency of deep molecular responses by 5 years was higher with nilotinib than with imatinib, besides, no patient in any arm has progressed to AP/BC on core treatment since the 2-year analysis. Progression to AP/BC after discontinuation of core treatment was reported in three patients between the 4-year and 5-year data cutoffs,5 including one patient in the nilotinib 300-mg twice-daily arm (progression reported ⩽28 days after discontinuation of nilotinib due to treatment failure (treatment-emergent T315I BCR-ABL mutation) and two patients in the imatinib arm (progression for both patients reported >3 years after discontinuation of imatinib due to Adverse event. (AE) (one patient) or treatment failure (treatment-emergent D276G BCR-ABL mutation; one patient). Regarding treatment-emergent BCR-ABL mutations were detected in 12/282 (4.3%) patients nilotinib 300-mg twice-daily, 11/281 (3.9%)%) patients, nilotinib 400-mg twice-daily and 22/283 (7.8%) and imatinib arms, including 4/282 (1.4%), 2/281 (0.7%) and 3/283 (1.1%) patients, respectively, with the BCR-ABL T315I mutation.. :and one in the imatinib arm (F317L) were detected by the 3-year data cutoff and were previously described in detail [198]: both of the patients with a treatment-emergent BCR-ABL mutation detected after the 3-year data cutoff had high Sokal risk scores at baseline. Adverse events (AEs) were reported using the Common Terminology Criteria for Adverse Events version 3.0.(4) (National Cancer Institute Common Termionology Criteria for Adverse Events (CTCAE) version 3.0.): the total frequencies of patients with grade 3/4 AEs, serious AEs and AEs leading to discontinuation of study treatment were comparable in the nilotinib 300-mg twice-daily (169 (60.6%), 72 (25.8%) and 34 (12.2%), respectively) and imatinib (165 (58.9%), 71 (25.4%) and 39 (13.9%), respectively) arms, and slightly higher in the nilotinib 400-mg twice-daily arm (198 (71.5%), 91 (32.9%) and 55 (19.9%), respectively). Furthermore, the most common nonhematologic AEs of any cause in both nilotinib arms were rash and headache; in the imatinib arm, diarrhea and nausea were most common. Moreover, in all three treatment arms, fluid retention, pleural effusion, pericardial effusion, pulmonary edema and cardiac tamponade (not reported in any arm)) was less common with nilotinib (300 mg twice daily, 11.1% 400 mg twice daily, 14.4%) than with imatinib (23.2%). Hypertension was reported in 10.4%, 8.3% and 4.3% of patients in the nilotinib 300-mg twice-daily, nilotinib 400-mg twice-daily, and imatinib arms, respectively.

Bosutinib (SKI-606, Bosulif®; Pfizer, New York, NY, USA) acts as a dual inhibitor of Src and ABL kinases. Compared with other second-generation tyrosine kinase inhibitors (TKIs), it has only minimal inhibitory activity against platelet-derived growth factor receptor -A or c-KIT.1,2 Bosutinib is a potent second-generation inhibitor of chronic myeloid leukemia (CML) cell proliferation in vitro and has been found to be capable of overcoming the majority of imatinib-resistant BCR-ABL mutations

[194]. BELA is an open-label, randomized, multinational, phase III study. Patients were randomly assigned 1:1 to bosutinib 500 mg per day or imatinib 400 mg per day. Patients could continue study treatment until disease progression or early discontinuation. Patients were discontinued from treatment if they could not tolerate at least 300 mg per day of bosutinib or imatinib. Dose escalation to 600 mg per day of either bosutinib or imatinib was allowed for patients not achieving an optimal response per current European LeukemiaNet guidelines [199], if no drug-related toxicity had been observed. The study of 502 patients with newly diagnosis of CP CML were enrolled between February 2008 and July 2009, in 139 centers in 31 countries; 250 patients were randomly assigned to bosutinib, and 252 patients were randomly assigned to imatinib. The current study was based on at least 12 months after the last patient's first visit, the median duration of treatment for both study arms was 13.8 months (bosutinib range, 0.03 to 30.3 months; imatinib range, 0.5 to 25.1 months). Median doses were 489 mg per day for bosutinib (range, 115 to 575 mg per day) and 400 mg per day for imatinib (range, 201 to 542 mg per day). During the study, nine patients (4%) receiving bosutinib and 30 patients (12%) receiving imatinib had their dose escalated to 600 mg per day because of lack of efficacy. Furthermore, as of the data cutoff date, 71% of patients in the bosutinib arm and 80% in the imatinib arm were still receiving study treatmen. Bosutinib did not demonstrate a superior rate of CCyR at 12 months versus imatinib based on the ITT [intent-to-treat) population, but it resulted in a superior rate of MMR at 12 months, which was the key secondary end point of the study. Furthermore, the authors have demostrated with this study that patients receiving bosutinib had a lower rate of disease progression/lack of efficacy (3%) compared with patients receiving imatinib (10%), including a trend toward fewer transformations to AP/BP CML (2% v 4%, respectively). these results indicate that bosutinib may be effective in some patients with newly diagnosed CP CML even though the primary efficacy end point was not met. Several authors have also been two recent studies with the second-generation TKIs dasatinib and nilotinib versus imatinib in patients with CP CML in the first-line setting [182, 193]. Cortes and others [200] have demostrated that efficacy in the intention-to-treat (ITT) population, with the

CCyR rate at 12 months was similar between patients receiving bosutinib (70%; 95% CI, 64% to 76%) and imatinib (68%; 95% CI, 62% to 74%; P = .601;), with 21 patients (4%; 10 patients on bosutinib; 11 on imatinib) included as having a CCyR at 12 months based on fluorescence in situ hybridization data only. Therefore, a superior CCyR rate at 12 months was not demonstrated for bosutinib versus imatinib, and the primary study end point was not met. However, median time to first CCyR was faster with bosutinib (12.9 weeks; 95% CI, 12.6 to 13.4 weeks) compared with imatinib (24.6 weeks; 95% CI, 24.3 to 25.6 weeks; P < .001), with higher rates of CCyR at months 3, 6, and 9 for bosutinib (Figure 2A). The cumulative rate of CCyR by 12 months was also similar between bosutinib and imatinib (79%; 95% CI, 74% to 84% v 75%; 95% CI, 69% to 80%, respectively). The major cytogenetic response rate at 12 months was 73% for bosutinib and 78% for imatinib. The MMR rate at 12 months was higher with bosutinib (41%; 95% CI, 35% to 47%) than imatinib (27%; 95% CI, 22% to 33%; P < .001, with the complete molecular response rate at 12 months being 12% compared with 3%, respectively (P < .001 Furthermore,the bosutinib was associated with a toxicity profile distinct from that of imatinib. In an unexpectedly high number of patients (n = 48, 19%) were discontinued from bosutinib treatment as a result of an AE, including 15 (31%) of 48 patients who discontinued before their first postbaseline assessment, infact early discontinuations as a result of toxicity may have contributed to the lower rate of CCyR associated with bosutinib in the ITT population, which included these patients as non- responders (82). The bosutinib therapy was associated with greater rates of diarrhea, vomiting, and aminotransferase elevations compared with imatinib, most instances were transient and, in the case of diarrhea, frequently self-limiting after the first 2 to 3 months of therapy. Fluid retention, which has commonly been observed with dasatinib [193, 201, 202] and imatinib [201-203] occurred infrequently with bosutinib. The occurrence of liver function test abnormalities with bosutinib treatment, but in no cases were associated with permanent liver injury. Low rates of grade 3 or 4 (all causality) anemia (6%), neutropenia (11%), and thrombocytopenia (14%) were reported with bosutinib. The minimal inhibition of c-KIT by bosutinib may have contributed to the low incidence of myelosuppressive events, particularly grade 3 or 4 neutropenia [204]. In conclusion, bosutinib did not achieve the primary efficacy end point of a superior rate of CCyR at 12 months versus imatinib; however, bosutinib did demonstrate other potential benefits in patients with newly diagnosed CP CML, in the end further follow-up is needed to provide long-term data on duration of response, transformation to AP/BP CML, and overall survival, as well as the tolerability profile of bosutinib in newly diagnosed CP CML.

Ponatinib is derived from AP23464, one of Ariad's dual ABL/SRC inhibitor, and is the first TKI demonstrating effectiveness against T315I mutation [205]. Ponatinib possesses a linear structure, which possibly helps it avoid steric clashes with hydrophobic gatekeeper residues and making it a potent inhibitor of most of know BCR-ABL mutations, particularly T315I [206]. Ponatinib is a third-generation tyrosine kinase inhibitor (TKI) with potent activity against native BCR-ABL1 and clinically relevant resistant mutants, including the BCR-ABL1^{T315I} gatekeeper mutant, which confers a high degree of resistance to all other currently available TKIs [206, 207]. Initially was granted accelerated approval by the U.S FDA in December 2012 for the treatment of patients with CML that were resistant to, or intolerant of prior TKI therapy, or Philadelphia chromosome–positive acute lymphoblastic leukemia (Ph+ ALL) and those harboring the resistant BCR-ABL1^{T315I} mutant [208, 209]. The pivotal phase 2 Ponatinib Ph+ ALL and CML Evaluation (PACE) trial evaluated the efficacy and safety of ponatinib at a starting dose of 45 mg once daily in CML or Ph+ ALL patients with resistance or intolerance to dasatinib or nilotinib, or with the BCR-ABL1^{T315I} mutation [210]. PACE was a phase 2 trial in adult patients with CML or Ph+ ALL who were resistant or intolerant to dasatinib or nilotinib, or who had the BCR-ABL1T315I mutation regardless of prior TKI use (N = 449; n = 270 CP-CML). The trial was recently completed and data as of the final database lock (6 February 2017) are reported at the time of study closure, patients who continued to derive clinical benefit from ponatinib had the option to continue therapy outside the clinical trial. Details of the study design and eligibility criteria have been previously published [210]. The starting dose for ponatinib was 45 mg once daily, and dose reductions to 30 mg or 15 mg once daily were applied to manage adverse events (AEs), per protocol, or implemented proactively following recommendations from the sponsor in October 2013 in response to concerns regarding an accumulation of AOEs with continued follow-up. Between September 2010 and October 2011, 449 patients were enrolled, including 270 CP-CML, 85 AP-CML, 62 BP-CML, and 32 Ph+ ALL patients, median age at baseline was 59 years (range, 18–94 years), and 47% were female. 93% patients had previously received at least 2 approved TKIs and 56% of patients had previously received at least 3 approved TKIs, and 29% of patients had the BCR-ABL1T315I mutation at study entry. Furthermore, the results presented here reflect data analysis as of 6 February 2017, with median follow-up of 37.3 months for all patients and 56.8 months (range, 0.1-73.1 months) for CP-CML patients. Time from the last enrolled patient's first visit to date of analysis was 64.1 months. It was noted that patients with more advanced disease experienced higher rates of discontinuation due to progressive disease and death. In conclusion, 04 patients (68%) had at least 1 dose reduction and 320 patients (71%) had a dose interruption of at least 3 days at any time; among CP-CML patients. Most patients with dose reductions already had their dose reduced prior to October 2013; as of 3 September 2013, 59% of patients (265 of 449), including 71% of CP-CML patients (192 of 270), had at least 1 dose reduction. In the initial report, 56% of CP-CML patients achieved the primary end point of major cytogenetic response (MCyR) by 12 months [210]. Besides, in studies of CP-CML patients who received their second TKI after resistance or intolerance to imatinib only: 57% of them achieved MCyR while receiving dasatinib (100 mg once daily), with 87% of these patients maintaining MCyR at 2 years [211] with 62% of these patients having MCyR lasting >18 months [212]; and 59% achieved or maintained baseline MCyR while receiving bosutinib, with a 77% probability of maintaining MCyR at 2 years. In the only other prospective evaluation of a TKI for patients treated with 2 or (rarely) 3 prior TKIs, 32% of CP-CML patients achieved MCyR with bosutinib, and the probability of maintaining MCyR at 2 years was 59% [213]. Responses in CP-CML patients in PACE have deepened over time, with 24% of CP-CML patients having achieved Minimal Response (MR) [210] at any time. Responses achieved by CP-CML patients in PACE correlated with favorable long-term outcomes, with estimated PFS/OS rates at 5 years of 53%, 73% overall, and with comparable rates in the resistant/intolerant and BCR-ABL1T315I cohorts. Furthermore, Long-term estimated OS rates among CP-CML patients with the BCR-ABL1T315I mutation, in particular, were in contrast to those that have been reported for any other approved TKI [214]. The multivariate analyses conducted to evaluate the impact of clinical factors on the response rates observed in the PACE study have shown that the presence of a T315I mutation is not an independent predictor of better response [215]. Rather, patients with T315I tended to have characteristics that are associated with higher response rates (ie, T315I patients were generally younger, more recently diagnosed, and treated with fewer prior TKIs, and they received higher doses of ponatinib [210]. In a separate analysis, no single or compound mutation was identified that consistently conferred primary and/or secondary resistance to ponatinib in CP-CML patients [216]. Therefore, ponatinib is effective in CP-CML patients irrespective of baseline mutation status. The continued follow-up of patients in the PACE trial has elucidated the vascular occlusive event profile in this population. Patients receiving ponatinib should be monitored for high blood pressure, evidence of arterial occlusive or thromboembolic events, and reduced cardiac function [208, 209]. These conditions should be managed as clinically indicated, and ponatinib dosing should be reduced, interrupted, or discontinued as needed [208, 209 217, 218]. Further research on the mechanisms of vascular events with TKIs is ongoing, and may help determine more specific interventions to mitigate or eliminate risk [219] (104) Instead ponatinib therapy does not appear to modify platelets [220] or the plasma lipid or the plasma lipid profile [221], although the mechanistic basis for ponatinib-associated AOEs is unknown, this vascular toxicity appears to be dose-related [222, 223] and modified by preexisting cardiovascular disease and other risk factors. In summary, these final results of the PACE study support ponatinib as an effective treatment of patients with CML who have received prior therapies, besides the decision to initiate ponatinib therapy should be guided by carefully weighing the risk-to-benefit ratio for each patient, especially in those who may be at increased risk of AOEs.

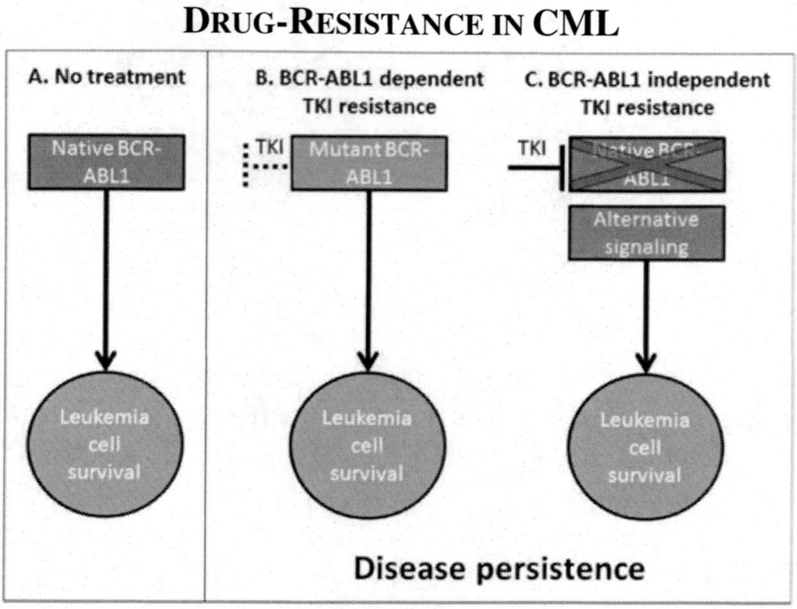

Figure 3. BCR-ABL1-dependent vs. independent resistance. (A) Native BCR-ABL1 signaling in the absence of TKI inhibition is necessary and sufficient for leukemogenesis in CML. (B) Kinase domain mutations in BCR-ABL1 can alter the binding of TKIs and lead to reconstitution of BCR-ABL1 signaling. (C) In the setting of effective BCR-ABL1 inhibition with TKIs, leukemia cells persist due to activation of alternative survival pathways (by article Ami B. Patel, Thomas O'Hare, and Michael W. Deininger, Mechanisms of resistance to ABL kinase inhibition in CML and the development of next generation ABL kinase inhibitors. Hematol Oncol Clin North Am. 2017 August; 31(4): 589–612).

Many authors have classified TKI resistance as either BCR-ABL1-dependent or BCR-ABL1 independent. This distinction seems formalistic, but it does have a great clinical relevance, as it informs the strategy required to combat resistance: BCR-ABL1-dependent resistance is reliant upon mechanisms that can subvert effective BCR-ABL1 kinase inhibition, such as point mutations in the kinase domain that impairs drug binding or cellular/biological processes that interferes with TKIs availability and results in suboptimal drug concentrations at the target. In contrast, BCR-ABL1-independent resistance is mediated through alternative survival pathways operating in the context of effective TKIs inhibition of BCR-ABL1 (Figure 3).

BCR-ABL1 Dependent

An analysis performed by Gorre et al. [163] clarified the mechanism of resistance in imatinib - treated patients in cases of relapsed blast crisis CML and Ph+ ALL, resistance was associated with active Bcr–Abl kinase despite continued imatinib treatment. Patient cells displayed an impaired sensitivity to imatinib ex vivo, demonstrating that resistance represented a cell-intrinsic phenomenon. In three out of 11 cases, amplification of the BCR–ABL gene was found. Furthermore, in one case of those patients, BCR–ABL amplification disappeared after imatinib suspension [163]. In addition, nine patients were analyzed for the presence of point mutations in BCR–ABL that might confer resistance by interfering with binding of imatinib at the kinase domain. Strikingly, a change of C to T at nucleotide 1091 leading to an exchange of a threonine at position 315 to isoleucine (T315I) was found in six out of nine cases. One patient had both, BCR–ABL amplification and mutation. Other studies demonstrated that a crystal structure analysis of the catalytic domain of Abl in complex to a variant form of imatinib predicted the presence of a threonine at position 315 to be a key requirement for binding with imatinib [224]. An exchange to isoleucine at this position prevents the formation of a critical hydrogen bond and adds an extra hydrocarbon group in its side chain, thereby leading to a steric clash with the drug [224] (Figure 4). Furthermore, von Bubnoff N and others [225] have stated with studies in vitro that that hematopoietic cell lines transduced with different mutant forms of BCR–ABL display resistance towards imatinib. Furthermore, the results of in vitro analysis of some mutant forms of Bcr–Abl (Y253H, E255K, E255V, and T315I) displayed IC50 values (concentrations where 50% of inhibition of biologic activities exhibited by Bcr–Abl occurs) that by far exceed concentrations measured in patients treated with imatinib [226, 227]. In contrast, the BCR–ABL mutant H396P had IC50 values that were nearby concentrations measured in treated patients [225, 226]. This result indicated that increasing dosage of imatinib may be beneficial in some cases of imatinib resistance due to mutated BCR–ABL. Corbin and others [228]. have demonstrated that T315I retained only

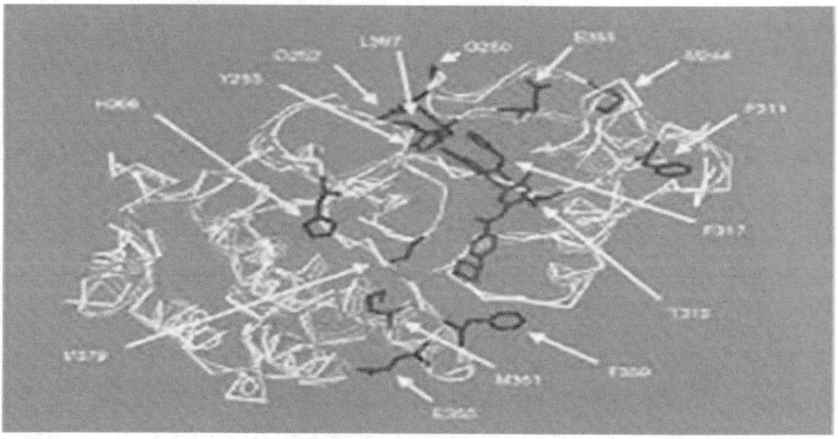

Figure 4. Positions of BCR–ABL gene mutations. Ribbon representation of the c-Abl kinase domain in complex with imatinib. The polypeptide backbone is shown in white, helices, sheets, and loops in pale pink, imatinib in green; side chains of mutated residues (M244, G250, Q252, Y253, E255, F311, T315, F317, M351, E355, F359, V379, L387, and H396) in red. Generated with PDB 1IEP available from http://www.rcsb.org/pdb/ (by article N von Bubnoff, C Peschel and J Duyster. Resistance of Philadelphia-chromosome positive leukemia towards the kinase inhibitor imatinib (STI571, Glivec): a targeted oncoprotein strikes back. Leukemia (2003) 17, 829–83).

61% of its kinase activity, suggesting that a possibly pre-existing clone harboring T315I may confer a growth advantage over wild-type Bcr–Abl only in the presence of imatinib, In fact, T315I displayed a two-fold increase in its ATP affinity relative to wild-type Abl and this situation may contribute to resistance to this and other mutations affecting the ATP binding site. Nilotinib overcomes resistance through tighter binding to a very similar (inactive) ABL1 conformation. Dasatinib was initially reported to bind to ABL1 with less stringent conformational requirements compared to imatinib, in fact the nuclear magnetic resonance studies suggested to be a type I inhibitor [229]. The dasatinib resistance mutation spectrum includes V299 and F317 as hotspots [230]. Furthermore, both nilotinib and dasatinib make a hydrogen bond with T315 and consequently have no activity against T315I. In vitro studies for nilotinib showed that mutations at positions Y253, E255, and T315 were consistently identified, while similar studies performed with dasatinib identified a small set of mutations that was essentially limited to contact points, including V299, T315, and F317 [231].

However, the percentage of patients positive for BCR-ABL1 KD mutations is approximately 30% in case of resistance to first line TKI treatment and rises up to 50–60% in case of resistance to second line TKI treatment [232]. Furthermore, In patients already harboring mutations selected by imatinib treatment, acquisition of new mutations conferring resistance to second-line therapy may give rise to very aggressive multi-mutated clones ('compound mutants') that are very difficult to counteract [233, 234]. The study, among the imatinib-resistant CML patients who switched to second-line TKIs therapy, was executed was executed BCR-ABL1 transcript level monitoring and KD mutation screening in a total of 125 peripheral blood samples. 51 later acquired dasatinib-(n = 26) or nilotinib-resistant mutations (n = 25) detected by conventional sequencing at the time of Failure, after a median of 9 months (range, 3–27 months) of therapy. Besides, DS reanalysis was performed from the time of failure and mutation detection by conventional sequencing backwards. Among the 26 patients who relapsed on dasatinib, 13 had acquired a T315I mutation, 10 had acquired F317L or V mutations, and 3 had acquired a V299L mutation (Figure 4). DS allowed to backtrack mutations in 11 cases (T315I, n = 2; F317L/V, n = 6; V299L, n = 3). In 2 patients, the mutations were detected at baseline. Furthermore, among the 25 patients who relapsed on nilotinib, 4 had acquired a T315I mutation, 8 had acquired an E255K or V mutation, 6 had acquired an F359V or I mutation, 1 had acquired an F359C and an E255K simultaneously, and 6 had acquired a Y253H mutation. DS allowed to backtrack mutations in 12 cases (T315I, n = 1; E255K/V, n = 6; F359V/I, n = 2; Y253H, n = 3). In 3 cases, the mutations were detected at baseline [In chronic myeloid leukemia patients on second-line tyrosine kinase inhibitor therapy, deep sequencing of BCR-ABL1 at the time of warning may allow sensitive detection of emerging drug-resistant mutants [235]. Redaelli S and others analyzed that bosutinib's resistance mutation spectrum is similar to that of dasatinib, suggesting that type I binding is dominant [236]. In contrast ponatinib is a type II inhibitor that binds ABL1 in a conformation that is quite similar to that observed with imatinib, except that no hydrogen bond is formed with T315 [206]. Moreover, ponatinib demonstrated activity in patients with or without BCR-ABL1 mutation (including T315I) present at baseline, as determined by

Sanger sequencing (SS). In addition, analysis of mutations in a subset of patients in the PACE trial who discontinued therapy (after a 15-month median follow-up) failed to identify any single mutations consistently associated with secondary resistance to ponatinib. However, in this and a separate study,5 certain compound mutations were associated with secondary resistance, though these were seen predominantly in patients with advanced CML or Ph+ ALL [210]. Several in vitro studies predict that a 30 mg/day dose would maintain efficacy against 7/8 non-T315I compound mutants tested in our panel. At a daily dose of 15 mg, ponatinib is predicted to pre-empt outgrowth of 5/8 non-T315I compound mutants in our panel, with the Y253H/E255V, E255V/V299L and F317L/F359V mutants remaining potentially problematic. In contrast, therapeutic utility is less promising with respect to T315I-inclusive compound mutants, where 9/10 in our panel showed little or no sensitivity to ponatinib or any of the other TKIs tested. Furthermore, some examples of clinical ponatinib failure attributable to E255V/T315I, T315I/F359C, Y253H/T315I, T315I/H396R, and T315I/E453K. Given the unique efficacy of ponatinib against the T315I single mutant and its current revised U.S. clinical indication, a significant fraction of future patients treated with ponatinib will be expected to harbor a T315I mutation at baseline. In addition, the broad potency of ponatinib against BCR-ABL1 point mutants can be traced to the extensive network of contacts that stabilize its binding to the kinase domain. In fact, certain pairings of mutations, each of which is susceptible to a given TKI in isolation, confer increased resistance when present as a compound mutation. For instance, several compound mutations involving the P-loop result in significant distortion of this region, suggesting it may prove advantageous to minimize direct TKI/P-loop interactions. Also intriguing is our characterization of an I315M point mutation in clinical resistance to ponatinib due to direct encroachment of the mutant side chain on drug binding (Patient #22). In fact, ponatinib is a poor inhibitor of kinases in which methionine is the native 'gatekeeper' position analogous to BCR-ABL1 position 315. For example, the insulin receptor is ~500 fold less sensitive to ponatinib than ABL1T315I [206]. By contrast, the T315A mutant is uniquely resistant to dasatinib and inhibited by each of the other five TKIs including ponatinib [173, 238]. These findings argue that efforts to develop future BCR-ABL1 TKIs should also consider the capacity to accommodate multiple different specific substitutions at the gatekeeper position. Other studies showed that failure to respond to TKI therapy could be due to alternative BCR-ABL1 independent mechanisms or BCR-ABL1-dependent mechanisms of resistance (such as efflux pumps). Such studies have demonstrated that the human organic cation transporter-1 (OCT-1) is the major active influx pump for imatinib. Low OCT-1 activity is associated with BCR-ABL1 dependent imatinib resistance. While high OCT-1 activity is correlated with improved MMR rates, in patients treated with imatinib [239, 240]. Furthermore, OCT-1 does not regulate cellular uptake of dasatinib, nilotinib or ponatinib [241], for this reason future researches are necessary to identify new candidates that regulate both imatinib monitoring and imatinib dose intensification.

BCR-ABL1 INDEPENDENT RESISTANCE

The causes of TKI resistance in primary resistance cases are more diverse and complicated. Donato et al, established an imatinib resistant K562 cell line (K562R) by treating an imatinib sensitive K562 cell line with increasing dose of imatinib, and found that overexpression of an SRC family kinase, Lyn, being the main driver of resistance [243]. This is further supported by the fact that K562R cells respond well to dasatinib, which inhibits both BCR-ABL and SRC family kinases [244]. Another recent work to identify genes involved in imatinib resistance using a genome-wide shRNA library revealed dozens of in-sensitizing genes [245]. Knocking down many of them using shRNA leads to imatinib resistance which is at least partially due to transcriptional upregulation of a PKC family member, PRKCH. Overexpression of PRKCH sustains the activity of MEK/ERK pathway upon IM treatment by directly phosphorylating and activating upstream CRAF. PRKCH/MAPK pathway was found to be a common resistance pathway in many BCR-ABL independent IM resistant cases. It is thought that BCR-ABL inhibition is the trigger of cell death due to quick loss of pro-survival signals while gaining pro-apoptotic signals [246]. However, cell signaling is always responding in a network fashion, and compensatory pathways may be transiently activated to slow down or circumvent cell death program [247], thereby causing TKI resistance. One of such examples is BCL6, which is an oncogene suppressed by JAK/STAT pathway. Upon TKI treatment, JAK/STAT pathway losses the activation signal from BCR-ABL, and releases the brake on BCL6, which then acts as an antiapoptotic Protein [248]. Several studies have demonstrated that while BCR-ABL1 independent resistance can confer overt resistance in active disease and it is also an important contributor to minimal residual disease (MRD), likely accounting for leukemia stem cell persistence despite deep molecular response (DMR) to TKI therapy. Multiple (and counting) signaling pathways have been implicated in BCR-ABL1-independent resistance. Furthermore, Eiring AM and others [242] demonstrated that resistance may enable survival of leukemogenic cells despite TKI inhibition of BCR-ABL1 through STAT3 activation. In experimental studies of co-culture of pSTAT3Y705 in TKI-naïve CML progenitor cells, grown in HS-5 bone marrow stromal cell-conditioned medium in the presence of IM, found that pSTAT3Y705 was able to promote CML cell survival. Such co-culture of TKI-sensitive CML primary cells with HS-5 human bone marrow stromal cells activated STAT3Y705 phosphorylation and leukemia cell survival through soluble BM-derived factors despite BCR-ABL1 inhibition. Furthermore, the authors have also found pSTAT3Y705 activation in CML CD34+ cells from patients with intrinsic TKI resistance (clinical resistance in the absence of BCR-ABL1 mutations). Other biological studies using TKI-resistant progenitors, showed that BP-5-087 (a highly potent and selective STAT3 inhibitor) combined with imatinib reduced survival and colony formation by CML progenitor and stem cells from patients with BCR-ABL1-independent resistance, consistent with a synthetically lethal pharmacologic interaction. Taken together, such data indicated that activation of pSTAT3Y705 was central to BCR-ABL1 kinase-independent TKI resistance [249, 250]. Furthermore, other studies found that activation of the PI3K/AKT/mTOR pathway was able to promote to facilitate primary CML cell survival during imatinib treatment until overt resistance through secondary mutations emerges [251]. Furthermore, a study showed that CML primary cells, treated with nilotinib and the PI3K inhibitor (NVP-BEZ235), underwent an inhibition of growth and increased apoptosis [252]. Increased cytoplasmic retention of Increased FOXO1 expression, (a transcription factor downstream of the PI3K signaling axis), can also contribute to BCR-ABL1 independent resistance in TKI-resistant CML cell lines [253]. MAP kinase signaling plays an important role in imatinib-treated CD34+ CML progenitor cells [254]. In line with these, another study [255] showed that Ras was activated in imatinib, nilotinib and dasatinib-resistant chronic myeloid leukemia cells, and consequently RAS was able to activate BRAF, C-RAF, MEK, and ERK, pathway in these cells Subsequently, nilotinib synergized with MEK inhibitors to kill drug-resistant CML cells and block tumor growth in mice. Besides, other authors found that imatinib activated ERK pathway in imatinib-resistant leukemia cells [256].

Many studies have implicated SphK1/S1P signalling in the process of drug resistance in both acute and chronic myelogenous leukemia. Elevated levels of SphK1 have been identified in variety of leukemic cell lines, correlating with chemotherapeutic resistance, while high SphK1 expression appears to be integral for erythroleukaemic progression [257]. Furthermore, SphK1 expression has been shown to be up regulated by BCR-ABL gene fusion, with this event necessary for Myeloid cell leukemia-1 (Mcl-1)-expressing and enhanced cell survival in CML. Recent studies elucidated the molecular mechanism by which SphK1 induces the acquisition of resistance to the anticancer agent imatinib in CML [257]. CML patients express the oncoprotein BCR-ABL, a constitutively active tyrosine kinase that activates multiple signaling pathways to promote cancer progression. Imatinib, a BCR-ABL inhibitor, is the mainstay therapeutic option for treatment of CML. SphK1 contributes to the resistance of CML cells to this anticancer agent as demonstrated by the finding that siRNA knock-down of SphK1 expression sensitizes resistant CML cells to imatinib whereas enforced expression of SphK1 prevents apoptosis in response to this drug in sensitive CML cells [259, 260]. Salas and colleagues [258] demonstrated that S1P/S1P2 signaling regulates BCR-ABL stability in CML cells by inhibiting the activity of the protein phosphatase PP2A. This prevents the dephosphorylation of BCR-ABL and therefore its proteasomal degradation, resulting in the accumulation of the oncoprotein. Thus, inhibition of the SphK1/S1P/S1P2 pathway sensitizes CML cells to imatinib by restoring the PP2A-dependent dephosphorylation and subsequent degradation of BCR-ABL [258]. Evidence also exists that the SphK1/ S1P/S1P receptors pathway enhances the expression and activity of drug efflux pumps in cancer cells, which are major contributors to the development of multidrug resistance in cancer.

CONCLUSION

New compounds and tools for the diagnosis and treatment of CRC and CML continue to be developed. Likewise, novel compounds and drug combinations offer the potential to delay or overcome acquired resistance to TKIs. C. Finally, it should be noted that there remains an unmet need to identify the cause of acquired resistance for numerous patients who stop responding to different TKs. Elucidation of resistance mechanisms to first- and second-generation TKIs should continue to be pursued as these TKIs are sure to retain clinical utility, and newly identified mechanisms will inform development of next-generation TKIs.

REFERENCES

[1] Bhandari A, Woodhouse M, Gupta S. Colorectal cancer is a leading cause of cancer incidence and mortality among adults younger than 50 years in the USA: a SEER-based analysis with comparison to other young-onset cancers. *J Investig Med.* 2017 Feb; 65(2):311-315.

[2] Gandomani HS, Yousefi SM, Aghajani M, Mohammadian-Hafshejani A, Tarazoj AA, Pouyesh V, Colorectal cancer in the world: incidence, mortality and risk factors Salehiniya H Colorectal cancer in the world:

incidence, mortality and risk factors. *Biomed Res Ther* 2017, 4(10): 1656-1675.

[3] Siegel RL, Miller KD, Jemal. A. Cancer statistics 2015 .CA: *Cancer Journal for. 2015; Clinicians* 65(1), 5-29.

[4] McKay JA, Loane JF, Ross VG, Ameyaw MM, Murray GI, Cassidy J, McLeod HL. c-erbB-2 is not a major factor in the development of colorectal cancer *Br J Cancer.* 2002 Feb 12; 86(4):568-73.

[5] Luo HY, Xu RH. Predictive and prognostic biomarkers with therapeutic targets in advanced colorectal cancer *World J Gastroenterol.* 2014 Apr 14; 20(14):3858-74.

[6] Arteaga, CL. The epidermal growth factor receptor: from mutant oncogenein non-human cancers to therapeutic target in human neoplasia. *J. Clin. Oncol.* 2001; 192 (Suppl. 18), 32S–40S.

[7] Venook, AP. Epidermal growth factor receptor-targeted treatment foradvanced colorectal carcinoma. *Cancer* 2005; 103, 2435–2446.

[8] Yokota T. Are KRAS/BRAF mutations potent prognostic and/or predictive biomarkers in colorectal cancers? *Anticancer Agents Med Chem.* 2012; 12(2):163-71.

[9] Colakoglu T, Yildirim S, Kayaselcuk F, Nursal TZ, Ezer A, Noyan T, Karakayali H, Haberal M. Clinicopathological significance of PTEN loss and the phosphoinositide 3-kinase/Akt pathway in sporadic colorectal neoplasms: is PTEN loss predictor of local recurrence? *Am J Surg.* 2008 195(6):719-25.

[10] Loewith R, Jacinto E, Wullschleger S, Lorberg A, Crespo JL, Bonenfant D, Bonenfant D, Oppliger W, Jenoe P, Hall MN. Two TOR complexes, only one of which is rapamycin sensitive, have distinct roles in cell growth control. *Mol Cell.* 2002; 10:457–468.

[11] Sarbassov DD, Ali SM, Kim DH, Guertin DA, Latek RR, Erdjument-Bromage H, Tempst P, Sabatini DM Rictor, a novel binding partner of mTOR, defines a rapamycin-insensitive and raptor-independent pathway that regulates the cytoskeleton. *Curr Biol.* 2004; 14(14):1296-302.

[12] Pause A, Belsham GJ, Gingras AC, Donzé O, Lin TA, Lawrence JC Jr, Sonenberg N. Insulin-dependent stimulation of protein synthesis

by phosphorylation of a regulator of 5'-cap function. *Nature,* 1994; 371, 762-7.

[13] Jastrzebski K, Hannan KM, Tchoubrieva EB, Hannan RD, Pearson RB. Coordinate regulation of ribosome biogenesis and function by the ribosomal protein S6 kinase, a key mediator of mTOR function. *Growth Factors,* 2007; 25, 209-26.

[14] Van Cutsem, E, Rougier, P, Köhne, C, Stroh, C, Schlichting, M, Bokemeyer, C. A meta-analysis of the CRYSTAL and OPUS studies combining cetuximab with chemotherapy (CT) as 1st-line treatment for patients (pts) withmetastatic colorectal cancer (mCRC): results according to KRAS and BRAFmutation status (abstr 6077). *Eur. J. Cancer.* 2009; Suppl. 7, 345.

[15] Cervantes, A, 2010. Advanced colorectal cancer: ESMO clinical practice guidelinesfor treatment. *Ann. Oncol.* 21 (Suppl 2), v93–v97.

[16] Bokemeyer, C, Khambata-Ford, S, Jonker, DJ, O'Callaghan, CJ, Tu, D, Tebbutt, NC, Simes, RJ, Chalchal, H, Shapiro, JD, Robitaille, S, Price, TJ, Shepherd, L, Au, HJ, Langer, C, Moore, MJ, Zalcberg, JR. Fluorouracil, leucovorin,and oxaliplatin with and without cetuximab in the first-line treatment of metastatic colorectal cancer. *J. Clin. Oncol.* 2009; 27, 663–667.

[17] Ciardiello, F, Bianco, R, Damiano, V, De Lorenzo, S, Pepe, S, De Placido, S, Fan, Z, Mendelsohn, J, Bianco, AR, Tortora, G. Antitumor activity of sequential treatment with topotecan and anti-epidermal growth factor receptormonoclonal antibody C225. *Clin. Cancer Res.* 1999. 5, 909–916.

[18] Cunningham, D, Humblet, Y, Siena, S, Khayat, D, Bleiberg, H, Santoro, A, Bets, D, Mueser, M, Harstrick, A, Verslype, C, Chau, I, Van, C, Van Custsem, E. Cetuximab monotherapy and cetuximab plus irinotecan inirinotecan-refractory metastatic colorectal cancer. *N. Engl. J. Med.* 2004; 351, 337–345.

[19] Saltz, LB, Meropol, NJ, Loehrer Sr, PJ, Needle, MN, Kopit, J, Mayer, RJ. Phase II trial of cetuximab in patients with refractory colorectal cancer thatexpresses the epidermal growth factor receptor. *J. Clin. Oncol.* 2004; 22, 1201–1208.

[20] Chung, KY, Shia, J, Kemeny, NE, Shah, M, Schwartz, GK, Tse, A, Hamilton, A, Pan, D, Schrag, D, Schwartz, L, Klimstra, DS, Fridman, D, Kelsen, DP, Saltz, LB. Cetuximab shows activity in colorectal cancer patients with tumorsthat do not express the epidermal growth factor receptor byimmunohistochemistry. *J. Clin. Oncol.* 2005. 23, 1803–1810.

[21] Lièvre, A, Bachet, JB, Boige, V, Cayre, A, Le, Corre, D, Buc, E, Ychou, M, Bouché, O, Landi, B, Louvet, C, André, T, Bibeau, F, Diebold, MD, Rougier, P, Ducreux, M, Tomasic, G, Emile, JF, Penault-Llorca, F, Laurent-Puig, P. KRAS mutations as an independent prognostic factor in patients with advancedcolorectal cancer treated with cetuximab. *J. Clin. Oncol.* 2008, 26, 374–379.

[22] Kohne, C, Stroiakovski, D, Chang-Chien, C, Lim, R, Pintér, T, Bodoky, G, Stroh, C, Celik, I, Rougier, P, Van Cutsem, EE, Oldenburg, K. Predictive biomarkers to improve treatment of metastatic colorectal cancer (mCRC): outcomes with cetuximab plus FOLFIRI in the CRYSTAL trial (abstr 4068). *J. Clin. Oncol.* 2009; 27 (Suppl).

[23] De Mattia, E, Cecchin, E, Toffoli, G. Pharmacogenomics of intrinsic andacquired pharmacoresistance in colorectal cancer: toward targeted personalized therapy. *Drug Resist.* Updat. 2015; pii: S1368-7646(15)00024-2.

[24] Di Nicolantonio, F, Martini, M, Molinari, F, Sartore-Bianchi, A, Arena, S, Saletti, P, De Dosso, S, Mazzucchelli, L, Frattini, M, Siena, S, Bardelli, A, Wild-type BRAF is required for response to panitumumab or cetuximab in metastaticcolorectal cancer. *J. Clin. Oncol.* 2008; 26, 5705–5712.

[25] Yonesaka, K, Zejnullahu, K, Okamoto, I, Satoh, T, Cappuzzo, F, Souglakos, J, Ercan, D, Rogers, A, Roncalli, M, Takeda, M, Fujisaka, Y, Philips, J, Shimizu, T, Maenishi, O, Cho, Y, Sun, J, Destro, A, Taira, K, Takeda, K, Okabe, T, Swanson, J, Itoh, H, Takada, M, Lifshits, E, Okuno, K, Engelman, JA, Shivdasani, RA, Nishio, K, Fukuoka, M, Varella-Garcia, M, Nakagawa, K, Jänne, PA. Activation of ERBB2 signaling causes resistance to the EGFR

directedtherapeutic antibody cetuximab. *Sci. Transl. Med.* 2011; 3, 99ra86.

[26] Normanno, N, Rachiglio, AM, Lambiase, M, Martinelli, E, Fenizia, F, Esposito, C, Roma, C, Troiani, T, Rizzi, D, Tatangelo, F, Botti, G, Maiello, E, Colucci, G, Ciardiello, F. On the behalf of the CAPRIGOIM investigators. Heterogeneity of KRAS, NRAS, BRAF and PIK3CA mutations in metastaticcolorectal cancer and potentialeffects on therapy in the CAPRI GOIM trial. *Ann Oncol.* 2015; Apr 7. pii: mdv1769.

[27] Ciardiello, F, Normanno, N, Maiello, EM, Martinelli, E, Troiani, T, Pisconti, S, Giuliani, F, Barone, C, Cartenì, G, Rachiglio, AM, Montesarchio, V, Tonini, G, Rizzi, D, Cinieri, S, Bordonaro, R, Febbraro, A, De, V, ita, F, Orditura, M, Fenizia, F, Lambiase, M, Rinaldi, A, Tatangelo, F, Botti, G, Colucci, G. Clinical activity of FOLFIRI plus cetuximab according to extended genemutation status by next-generation sequencing: findings from the CAPRI-GOIM trial. *Ann. Oncol.* 2014; 25, 1756–1761.

[28] Martinelli E, Cardone C, Troiani T, Normanno N, Pisconti S, Sforza V, Bordonaro AR, Rachiglio AM, Lambiase M, Latiano TP, Modoni G, Cordio S, Giuliani F, Biglietto M, Montesarchio V, Barone C, Tonini G, Cinieri S, Febbraro A, Rizzi D, De Vita F, Orditura M, Colucci G, Maiello E, Ciardiello F. Clinical activity and tolerability of FOLFIRI and cetuximab in elderly patients with metastatic colorectal cancer in the CAPRI-GOIM first-line trial. *ESMO Open.* 2017 Feb 20;1(6): e000086. doi: 10.1136/esmoopen-2016-000086. eCollection 2016.

[29] Normanno N, Esposito Abate R, Lambiase M, Forgione L, Cardone C, Iannaccone A, Sacco A, Rachiglio AM, Martinelli E, Rizzi D, Pisconti S, Biglietto M, Bordonaro R, Troiani T, Latiano TP, Giuliani F, Leo S, Rinaldi A, Maiello E, Ciardiello F; CAPRI-GOIM Investigators. RAS testing of liquid biopsy correlates with the outcome of metastatic colorectal cancer patients treated with first-line FOLFIRI plus cetuximab in the CAPRI-GOIM trial. *Ann Oncol.* 2018 29:112-118.

[30] Taieb J, Balogoun R, Le Malicot K, Tabernero J, Mini E, Folprecht G, Van Laethem JL, Emile JF, Mulot C, Fratté S, Levaché CB, Saban-Roche L, Thaler J, Petersen LN, Bridgewater J, Perkins G, Lepage C, Van Cutsem E, Zaanan A, Laurent-Puig P; PETACC8 Investigators. Adjuvant FOLFOX +/- cetuximab in full RAS and BRAF wildtype stage III colon cancer patients. *Ann Oncol.* 2017; 28(4):824-830.

[31] Yang YF, Wang GY, He JL, Wu FP, Zhang YN. Overall survival of patients with KRAS wild-type tumor treated with FOLFOX/FORFIRI±cetuximab as the first-line treatment for metastatic colorectal cancer: A meta-analysis. *Medicine* (Baltimore). 2017; 96(12): e6335.

[32] Boisdron-Celle M, Metges JP, Capitain O, Adenis A, Raoul JL, Lecomte T, Lam YH, Faroux R, Masliah C, Poirier AL, Berger V, Morel A, Gamelin E. A multicenter phase II study of personalized FOLFIRI-cetuximab for safe dose intensification. *Semin Oncol.* 2017; 44(1):24-33.

[33] Kotake M, Aoyama T, Munemoto Y, Doden K, Kataoka M, Kobayashi K, Nishimura G, Fujita H, Nakamura K, Takehara A, Tanaka C, Sakamoto J, Nagata N, Oba K, Kondo K. Multicenter phase II study of infusional 5-fluorouracil (5-FU), leucovorin, and oxaliplatin, plus biweekly cetuximab as first-line treatment in patients with metastatic colorectal cancer (CELINE trial). *Oncol Lett.* 2017; 13(2):747-753.

[34] Iwamoto S, Maeda H, Hazama S, Oba K, Okayama N, Suehiro Y, Yamasaki T, Suzuki N, Nagano H, Sakamoto J, Mishima H, Nagata N. Efficacy of CapeOX plus Cetuximab Treatment as a First-Line Therapy for Patients with Extended RAS/BRAF/PIK3CA Wild-Type Advanced or Metastatic Colorectal Cancer. *J Cancer.* 2018; 18; 9(22):4092-4098.

[35] Shapiro JD, Thavaneswaran S, Underhill CR, Robledo KP, Karapetis CS, Day FL, Nott LM, Jefford M, Chantrill LA, Pavlakis N, Tebbutt NC, Price TJ, Khasraw M, Van Hazel GA, Waring PM, Tejpar S, Simes J, Gebski VJ, Desai J, Segelov E. Cetuximab Alone or With Irinotecan for Resistant KRAS-NRAS-, BRAF- and PIK3CA-wild-

type Metastatic Colorectal Cancer: The AGITG Randomized Phase II ICECREAM Study. *Clin Colorectal Cancer.* 2018; 17(4):313-319.

[36] Aranda E, García-Alfonso P, Benavides M, Sánchez Ruiz A, Guillén-Ponce C, Safont MJ, Alcaide J, Gómez A, López R, Manzano JL, Méndez Ureña M, Sastre J, Rivera F, Grávalos C, García T, Martín-Valadés JI, Falcó E, Navalón M, González Flores E, Ma García Tapiador A, Ma López Muñoz A, Barrajón E, Reboredo M, García Teijido P, Viudez A, Cárdenas N, Díaz-Rubio E; Spanish Cooperative Group for the Treatment of Digestive Tumours (TTD). First-line mFOLFOX plus cetuximab followed by mFOLFOX plus cetuximab or single-agent cetuximab as maintenance therapy in patients with metastatic colorectal cancer: Phase II randomised MACRO2 TTD study Eur *J Cancer.* 2018; 101:263-272.

[37] Kienle DL, Dietrich D, Ribi K, Wicki A, Quagliata L, Winterhalder RC, Koeberle D, Horber D, Bastian S, Kueng M, Saletti P, Helbling D, Baertschi D, Lugli A, Bernhard J, Andrieu C, von Moos R. Cetuximab monotherapy and cetuximab plus capecitabine as first-line treatment in older patients with RAS- and BRAF wild-type metastatic colorectal cancer. Results of the multicenter phase II trial SAKK 41/10. *J Geriatr Oncol.* 2019; 10(2):304-310.

[38] Toledo RA, Cubillo A, Vega E, Garralda E, Alvarez R, de la Varga LU, Pascual JR, Sánchez G, Sarno F, Prieto SH, Perea S, Lopéz-Casas PP, López-Ríos F, Hidalgo M. Clinical validation of prospective liquid biopsy monitoring in patients with wild-type RAS metastatic colorectal cancer treated with FOLFIRI-cetuximab *Oncotarget.* 2017; 8(21):35289-35300.

[39] Qin S, Li J, Wang L, Xu J, Cheng Y, Bai Y, Li W, Xu N, Lin LZ, Wu Q, Li Y, Yang J, Pan H, Ouyang X, Qiu W, Wu K, Xiong J, Dai G, Liang H, Hu C, Zhang J, Tao M, Yao Q, Wang J, Chen J, Eggleton SP, Liu T. Efficacy and Tolerability of First-Line Cetuximab Plus Leucovorin, Fluorouracil, and Oxaliplatin (FOLFOX-4) Versus FOLFOX-4 in Patients With RAS Wild-Type Metastatic Colorectal: The Open-Label, Randomized, Phase III TAILOR Trial. *Cancer: J Clin Oncol.* 2018: JCO2018783183.

[40] Tanioka H, Asano M, Yoshida R, Waki N, Uno F, Ishizaki M, Yamashita K, Morishita Y, Nagasaka T. Cetuximab retreatment in patients with metastatic colorectal cancer who exhibited a clinical benefit in response to prior cetuximab: A retrospective study. *Oncol Lett.* 2018; 16(3):3674-3680. doi: 10.3892/ol.2018.9127.

[41] Pugh S, Thiébaut R, Bridgewater J, Grisoni ML, Moutasim K, Rousseau F, Thomas GJ, Griffiths G, Liebaert F, Primrose J, Laurent-Puig P. Association between miR-31-3p expression and cetuximab efficacy in patients with KRAS wild-type metastatic colorectal cancer: a post-hoc analysis of the New EPOC trial. *Oncotarget.* 2017; 8(55):93856-93866.

[42] Guo GF, Wang YX, Zhang YJ, Chen XX, Lu JB, Wang HH, Jiang C, Qiu HQ, Xia LP. Predictive and prognostic implications of 4E-BP1, Beclin-1, and LC3 for cetuximab treatment combined with chemotherapy in advanced colorectal cancer with wild-type KRAS: Analysis from real-world data. *World J Gastroenterol.* 2019; 25(15):1840-1853.

[43] Ledys F, Derangère V, Réda M, Guion JF, Milliex R, Roux V, Limagne E, Arnould L, Bengrine L, Ghiringhelli F, Rébé C. Anti-MEK and Anti-EGFR mAbs in RAS-Mutant Metastatic Colorectal Cancer: *Case Series and Rationale Adv Ther.* 2019; 36(6):1480-1484.

[44] Parseghian CM, Parikh NU, Wu JY, Jiang ZQ, Henderson L, Tian F, Pastor B, Ychou M, Raghav K, Dasari A, Fogelman DR, Katsiampoura AD, Menter DG, Wolff RA, Eng C, Overman MJ, Thierry AR, Gallick GE, Kopetz S. Dual Inhibition of EGFR and c-Src by Cetuximab and Dasatinib Combined with FOLFOX Chemotherapy in Patients with Metastatic Colorectal Cancer. *lin Cancer Res.* 2017; 23(15):4146-4154.

[45] Weber J, McCormack PL. Panitumumab: in metastatic colorectal cancer with wild-type KRAS. *Bio Drugs.* 2008; 22(6):403-11.

[46] Amado RG, Wolf M, Peeters M, Van Cutsem E, Siena S, Freeman DJ, Juan T, Sikorski R, Suggs S, Radinsky R, Patterson SD, Chang DD Wild-type KRAS is required for panitumumab efficacy in patients

with metastatic colorectal cancer. *J Clin Oncol.* 2008; 26(10):1626-34.

[47] Sobrero, AF, Peeters, M, Price, TJ, Hotko, Y, Cervantes-Ruiperez, A, Ducreux, M, Andre, T, Chan, E, Lordick, F, Tian, Y, Sidhu, R, *Final results of study 20,050,181: a randomized phase III study of FOLFIRI with our without panitumumab (pmab) for the second-line treatment (tx) of metastaticcolorectal cancer (mCRC),* poster 2012 ASCO annual meeting.

[48] Boeckx N, Koukakis R, Op de Beeck K, Rolfo C, Van Camp G, Siena S, Tabernero J, Douillard JY, André T, Peeters M. Primary tumor sidedness has an impact on prognosis and treatment outcome in metastatic colorectal cancer: results from two randomized first-line panitumumab studies. *Ann Oncol.* 2017; 28(8):1862-1868.

[49] Leone F, Artale S, Marino D, Cagnazzo C, Cascinu S, Pinto C, Fornarini G, Tampellini M, Di Fabio F, Sartore-Bianchi A, De Carlis L, Pugliese R, Capussotti L, Gioeni L, Siena S, Aglietta M. Panitumumab in combination with infusional oxaliplatin and oral capecitabine for conversion therapy in patients with colon cancer and advanced liver metastases. *The MetaPan study Cancer.* 2013; 119(19):3429-35.

[50] Munemoto Y, Kanda M, Oba K, Kim HM, Takemoto H, Denda T, Nagata N, Takano N, Fukunaga M, Kataoka M, Tokunaga Y, Sakamoto J, Mishima H. A phase II trial to evaluate the efficacy of panitumumab combined with fluorouracil-based chemotherapy for metastatic colorectal cancer: the PF trial. *Cancer Chemother Pharmacol.* 2018; 81(5):829-838.

[51] Boeckx N, Koukakis R, Op de Beeck K, Rolfo C, Van Camp G, Siena S, Tabernero J, Douillard JY, André T, Peeters M. Effect of Primary Tumor Location on Second- or Later-line Treatment Outcomes in Patients With RAS Wild-type Metastatic Colorectal Cancer and All Treatment Lines in Patients With RAS Mutations in Four Randomized Panitumumab Studies. *Clin Colorectal Cancer.* 2018; 17(3):170-178.e3.

[52] Kim TW, Elme A, Park JO, Udrea AA, Kim SY, Ahn JB, Valencia RV, Krishnan S, Manojlovic N, Guan X, Lofton-Day C, Jung AS, Vrdoljak E. Final Analysis of Outcomes and RAS/BRAF Status in a Randomized Phase 3 Study of Panitumumab and Best Supportive Care in Chemorefractory Wild Type KRAS Metastatic Colorectal Cancer. *Clin Colorectal Cancer.* 2018; 17(3):206-214.

[53] Papaxoinis G, Kotoula V, Giannoulatou E, Koliou GA, Karavasilis V, Lakis S, Koureas A, Bobos M, Chalaralambous E, Daskalaki E, Chatzopoulos K, Tsironis G, Pazarli E, Chrisafi S, Samantas E, Kaklamanos IG, Varthalitis I, Konstantara A, Syrigos KN, Pentheroudakis G, Pectasides D, Fountzilas G. Phase II study of panitumumab combined with capecitabine and oxaliplatin as first-line treatment in metastatic colorectal cancer patients: clinical results including extended tumor genotyping. *Med Oncol.* 2018; 35(7):101.

[54] Chen D, Li L, Zhang X, Gao G, Shen L, Hu J, Yang M, Liu B, Qian X. FOLFOX plus anti-epidermal growth factor receptor (EGFR) monoclonal antibody (mAb) is an effective first-line treatment for patients with RAS-wild left-sided metastatic colorectal cancer: A meta analysis. *Medicine* (Baltimore). 2018; 97(10):e0097.

[55] Borelli B, Moretto R, Lonardi S, Bonetti A, Antoniotti C, Pietrantonio F, Masi G, Burgio V, Marmorino F, Salvatore L, Rossini D, Zaniboni A, Zucchelli G, Martignetti A, Di Battista M, Pella N, Passardi A, Boccaccino A, Leone F, Colombo C, Granetto C, Vannini F, Marsico VA, Martinelli E, Antonuzzo L, Vitello S, Delliponti L, Boni L, Cremolini C, Falcone A. TRIPLETE: a randomised phase III study of modified FOLFOXIRI plus panitumumab versus mFOLFOX6 plus panitumumab as initial therapy for patients with unresectable *RAS* and *BRAF*wild-type metastatic colorectal cancer. *ESMO Open.* 2018; 3(4):e000403.

[56] Alonso V, Escudero P, Fernández-Martos C, Salud A, Méndez M, Gallego J, Rodriguez JR, Martín-Richard M, Fernández-Plana J, Manzano H, Méndez JC, Zanui M, Falcó E, Gil-Raga M, Rojo F, Cuatrecasas M, Feliu J, García-Albéniz X, Maurel J. *Modulates*

Panitumumab and Cetuximab Efficacy in RAS Wild-Type Metastatic Colorectal Cancer Patients Neoplasia. 2018; 20(7):678-686.

[57] Merx K, Martens UM, Kripp M, Hoehler T, Geissler M, Gaiser T, Mai S, Kienle P, Belle S, Plöger C, Hieber U, Wenz F, Post S, Hofheinz RD. Panitumumab in Combination With Preoperative Radiation Therapy in Patients With Locally Advanced RAS Wild-type Rectal Cancer: Results of the Multicenter Explorative Single-ArmPhase 2 Study NEORIT. *Int J Radiat Oncol Biol Phys.* 2017; 15; 99(4):867-875.

[58] Thatcher, N, Chang, A, Parikh, P, Rodrigues, P, Pereira, J, Ciuleanu, T, von, Pawel, J, Thongprasert, S, Tan, EH, Pemberton, K, Archer, V, Carroll, K. Gefitinib plus best supportive care in previously treated patients withrefractory advanced non-small-cell lung cancer: results from a randomised,placebo-controlled, multicentre study (Iressa Survival Evaluation in Lung Cancer). *Lancet* 2005; 366, 1527–1537.

[59] Wolpin, BM, Clark, JW, Meyerhardt, JA, Earle, CC, Ryan, DP, Enzinger, PC, Zhu, AX, Blaszkowsky, L, Battu, S, Fuchs, CS. Phase I. study of gefitinib plusFOLFIRI in previously untreated patients with meta- static colorectal cancer. *Clin. Colorectal Cancer* 2006; 6, 208–213.

[60] Chau, I, Cunningham, D, Hickish, T, Massey, A, Higgins, L, Osborne, R, Botwood, N, Swaisland, A. Gefitinib and irinotecan in patients withfluoropyrimidine-refractory, irinotecan-naive advanced colorectal cancer: aphase I-II study. *Ann. Oncol.* 2007; 18, 730–737.

[61] Eberhard, DA, Johnson, BE, Amler, LC, Goddard, AD, Heldens, SL, Herbst, RS, Ince, WL, Jänne, PA, Januario, T, Johnson, DH, Klein, P, Miller, VA, Ostland, MA, Ramies, DA, Sebisanovic, D, Stinson, JA, Zhang, YR, Seshagiri, S, Hillan, KJ. Mutations in the epidermal growth factor receptor and in KRAS arepredictive and prognostic indicators in patients with non-small-cell lungcancer treated with chemotherapy alone and in combination with erlotinib. *J. Clin. Oncol.* 2005; 23, 5900–5909.

[62] Paez, JG, Jänne, PA, Lee, JC, Tracy, S, Greulich, H, Gabriel, S, Herman, P, Kaye, FJ, Lindeman, N, Boggon, TJ, Naoki, K, Sasaki, H,

Fujii, Y, Eck, MJ, Sellers, WR, Johnson, BE, Meyerson, M. EGFR mutations in lung cancer:correlation with clinical response to gefitinib therapy. *Science* 2004; 304, 1497–1500.

[63] Sakurada, A, Shepherd, FA, Tsao, MS. Epidermal growth factor receptortyrosine kinase inhibitors in lung cancer: impact of primary or secondarymutations. *Clin. Lung Cancer* 7 2006; (Suppl. 4), S138–S144.

[64] Mitsudomi, T, Kosaka, T, Endoh, H, Horio, Y, Hida, T, Mori, S, Hatooka, S, Shinoda, M, Takahashi, T, Yatabe, Y, Mutations of the epidermal growth factor receptor gene predict prolonged survival after gefitinib treatment inpatients with non-small-cell lung cancer with postoperative recurrence. *J. Clin. Oncol.* 2005; 23, 2513–2520.

[65] Lee, CK, Wu, YL, Ding, PN, Lord, SJ, Inoue, A, Zhou, C, Mitsudomi, TR, Rosell, Pavlakis, N, Links, M, Gralla, RJ, Yang, JCH. Impact of specific Epidermal Growth Factor Receptor (EGFR) mutations and clinical characteristics on outcomes after treatment with EGFR tyrosine kinase inhibitors versus chemotherapy in EGFR-mutant lung cancer: A Meta-Analysis Epidermal growth factor receptor mutations in lung Adenocarcinoma *CO JCO*. 2015; 2014.58.1736; published online on April 20.

[66] Shigematsu, H, Lin, L, Takahashi, T, Nomura, M, Suzuki, M, Wistuba, II, Fong, KM, Lee, H, Toyooka, S, Shimizu, N, Fujisawa, T, Feng, Z, Roth, JA, Herz, J. Minna, JD, Gazdar, AF. Clinical and biological features associated withepidermal growth factor receptor gene mutations in lung cancers. *J. Natl. Cancer Inst.* 2005; 97, 339–346.

[67] Moroni, M, Veronese, S, Benvenuti, S, Marrapese, G, Sartore-Bianchi, A, DiNicolantonio, F, Gambacorta, M, Siena, S, Bardelli, A. Gene copy number for epidermal growth factor receptor (EGFR) and clinical response to anti EGFR treatment in colorectal cancer: a cohort study. *Lancet Oncol.* 2005; 6, 279–286.

[68] Nagahara, H, Mimori, K, Ohta, M, Utsunomiya, T, Inoue, H, Barnard, GF, Ohira, M, Hirakawa, K, Mori, M. Somatic mutations of epidermal

growth factor receptor in colorectal carcinoma. *Clin. Cancer Res.* 2005; 11, 1368–1371.

[69] Yunxia, Z, Jun, C, Guanshan, Z, Yachao, L, Xueke, Z, Jin, L. Mutations in epidermal growth factor receptor and K-ras in Chinese patients with colonrectal cancer. *BMC Med. Genet.* 2010:11, 34, 34.

[70] Barber, TD, Vogelstein, B, Kinzler, KW, Velculescu, VE. Somatic mutations of EGFR in colorectal cancers and glioblastomas. *N. Engl. J. Med.* 2004; 351, 2883.

[71] Metzger, B, Chambeau, L, Begon, DY, Faber, C, Kayser, J, Berchem, G, Pauly, M, Boniver, J, Delvenne, P, Dicato, M, Wenner, T. The human epidermal growth factor receptor (EGFR) gene in European patients with advanced colorectal cancer harbors infrequent mutations in its tyrosine kinase domain. *BMC Med. Genet.* 2011; 12, 144.

[72] Gridelli, C, Bareschino, MA, Schettino, C, Rossi, A, Maione, P, Ciardiello, F. Erlotinib in non-small cell lung cancer treatment: current status and futuredevelopment. *Oncologist* 2007; 12, 840–849.

[73] Meyerhardt, JA, Stuart, K, Fuchs, CS, Zhu, AX, Earle, CC, Bhargava, P, Blaszkowsky, L, Enzinger, P, Mayer, RJ, Battu, S, Lawrence, C, Ryan, DP. Phase II study of FOLFOX, bevacizumab and erlotinib as first-line therapy for patients with metastatic colorectal cancer. *Ann. Oncol.* 2007; 18, 1185–1189.

[74] Bajetta, E, Di Bartolomeo, M, Buzzoni, R, Ferrario, E, Dotti, KF, Mariani, L, Bajetta, R, Gevorgyan, A, Venturino, P, Galassi, M. Dose finding study oferlotinib combined to capecitabine and irinotecan in pretreated advancedcolorectal cancer patients. *Cancer Chemother. Pharmacol.* 2009; 64, 67–72.

[75] Munoz, A, Pericay, C, García-Girón, C, Alonso, V, Due˜nas, R, Cirera, L, Rivera, F, Falcó, E, Bustos, IA, Salud, AP. Phase II study of bevacizumab,capecitabine, and oxaliplatin followed by bevacizumab plus erlotinib asfirst-line therapy in metastatic colorectal cancer. *Oncol. Res.* 2013; 21, 181–191.

[76] Blaszkowsky, LS, Ryan, DP, Szymonifka, J, Borger, DR, Zhu, AX, Clark, JW, Kwak, EL, Mamon, HJ, Allen, JN, Vasudev, E, Shellito, PC, Cusack, JC, Berger, DL, Hong, TS, Phase I/II study of

neoadjuvant bevacizumab,erlotinib and 5-fluorouracil with concurrent external beam radiation therapyin locally advanced rectal cancer. *Ann. Oncol.* 2014, 121–126.

[77] Falchook, GS, Naing, A, Wheler, JJ, Tsimberidou, AM, Zinner, R, Hong, DS, Fu, S, Piha-Paul, SA, Janku, F, Hess, KR, Bastida, C, Kurzrock, R, Dual, E. FR inhibition in combination with anti-VEGF treatment in colorectal cancer. *Oncoscience* 2014; 1, 540–549.

[78] Carlomagno, C, Daniele, G, Bianco, R, Marciano, R, Damiano, V, Matano, E, Nappi, L, Pepe, S, De Placido, S, Tortora, G. Addition of erlotinib tofluoropyrimidine-oxaliplatin based chemotherapy with or without bevacizumab: two sequential phase I trials. *Exp. Ther. Med.* 2011; 2, 449–455.

[79] Weickhardt AJ, Price TJ, Chong G, Gebski V, Pavlakis N, Johns TG, Azad A, Skrinos E, Fluck K, Dobrovic A, Salemi R, Scott AM, Mariadason JM, Tebbutt NC. Dual targeting of the epidermal growth factor receptor using the combination of cetuximab and erlotinib: preclinical evaluation and results of the phase II DUX study in chemotherapy-refractory, advanced colorectal cancer. *J Clin Oncol.* 2012; 30(13):1505-12.

[80] Vincent MD, Breadner D, Soulieres D, Kerr IG, Sanatani M, Kocha W, Klimo P, MacKenzie MJ, O'Connell A, Whiston F, Malpage AS, Stitt L, Welch SA. Phase II trial of capecitabine plus erlotinib versus capecitabine alone in patients with advanced colorectal cancer. *Future Oncol.* 2017; 13(9):777-786.

[81] Kearney MR, Chen EY, Vaccaro GM, Strother J, Burt A, Todd K, Donovan J, Kampa-Schittenhelm KM, Lopez CD. Second-line mFOLFOX6 or FOLFIRI for Metastatic Colorectal Cancer. *Anticancer Res.* 2019; 39(1):245-252.

[82] Greally M, Kelly CM, Cercek A. HER2: An emerging target in colorectal cancer. *Curr Probl Cancer.* 2018; 42(6):560-571.

[83] Conradi LC, Styczen H, Sprenger T, Wolff HA, Rödel C, Nietert M, Homayounfar K, Gaedcke J, Kitz J, Talaulicar R, Becker H, Ghadimi M, Middel P, Beissbarth T, Rüschoff J, Liersch T. Frequency of HER-

2 positivity in rectal cancer and prognosis. *Am J Surg Pathol.* 2013; 37:522–531.

[84] Park DI, Kang MS, Oh SJ, Kim HJ, Cho YK, Sohn CI, Jeon WK, Kim BI, Han WK, Kim H, Ryu SH, Sepulveda AR. HER-2/neu overexpression is an independent prognostic factor in colorectal cancer. *Int J Colorectal Dis.* 2007; 22(5):491-7.

[85] Valtorta E, Martino C, Sartore-Bianchi A, Penaullt-Llorca F, Viale G, Risio M, Rugge M, Grigioni W, Bencardino K, Lonardi S, Zagonel V, Leone F, Noe J, Ciardiello F, Pinto C, Labianca R, Mosconi S, Graiff C, Aprile G, Frau B, Garufi C, Loupakis F, Racca P, Tonini G, Lauricella C, Veronese S, Truini M, Siena S, Marsoni S, Gambacorta M. Assessment of a HER2 scoring system for colorectal cancer: results from a validation study. *Mod Pathol.* 2015; 28(11):1481-91.

[86] Raghav KPS, Overman MJ, Yu R, Meric-Bernstam F, Menter D, Kee BK, Muranyi A, Singh S, Routbort M Chen K, Shaw KR, Shanmugam K, Maru DM, Fakih M Kopetz S HER2 amplification as a negative predictive biomarker for anti-epidermal growth factor receptor antibody therapy in metastatic colorectal cancer. *J Clin Oncol.* 2016; 34 3517–3517.

[87] Raghav, Kanwal & Loree, Jonathan & Morris, Jeffrey & Overman, Michael & Yu, Ruoxi & Meric-Bernstam, Funda & Menter, David & Korphaisarn, Krittiya & Kee, Brian & Muranyi, Andrea & Singh, Shalini & Routbort, Mark & Chen, Ken & R. M. Shaw, Kenna & Katkhuda, Riham & Shanmugam, Kandavel & Maru, Dipen & Fakih, Marwan & Kopetz, Scott. (2019). Validation of HER2 Amplification as a Predictive Biomarker for Anti Epidermal Growth Factor Receptor Antibody Therapy in Metastatic Colorectal Cancer. *JCO Precision Oncology.* 1-13. 10.1200/PO.18.00226.

[88] Richman SD, Southward K, Chambers P, Cross D, Barrett J, Hemmings G, Taylor M, Wood H, Hutchins G, Foster JM, Oumie A, Spink KG, Brown SR, Jones M, Kerr D, Handley K, Gray R, Seymour M, Quirke P. HER2 overexpression and amplification as a potential therapeutic target in colorectal cancer: analysis of 3256 patients

enrolled in the QUASAR, FOCUS and PICCOLO colorectal cancer trials. *J Pathol.* 2016; 238(4):562-70.

[89] Tu J, Yu Y, Liu W, Chen S. Significance of human epidermal growth factor receptor 2 expression in colorectal cancer. *Exp Ther Med* 2015; 9: 17-24.

[90] Ramanathan RK, Hwang JJ, Zamboni WC, Sinicrope FA, Safran H, Wong MK, Earle M, Brufsky A, Evans T, Troetschel M, Walko C, Day R, Chen HX, Finkelstein S. Low overexpression of HER-2/neu in advanced colorectal cancer limits the usefulness of trastuzumab (Herceptin) and irinotecan as therapy. A phase II trial. *Cancer Invest* 2004; 22: 858-865.

[91] Clark JW, Niedzwiecki D, Hollis D, Mayer R. Phase II trial of 5-fluorouracil (5-FU), leucovorin (LV), oxaliplatin (Ox), and trastuzumab (T) for patients with metastatic colorectal cancer (CRC) refractory to initial therapy. *Proc Am Soc Clin Oncol* 2003; 22: Abstract 35.

[92] Rubinson DA, Hochster HS, Ryan DP, Wolpin BM, McCleary NJ, Abrams TA, Chan JA, Iqbal S, Lenz HJ, Lim D, Rose J, Bekaii- Saab T, Chen HX, Fuchs CS, Ng K. Multi-drug inhibition of the HER pathway in metastatic colorectal cancer: results of a phase I study of pertuzumab plus cetuximab in cetuximab-refractory patients. *Invest New Drugs* 2014; 32: 113-122.

[93] Frank D, Jumonville A, Loconte NK, Schelman WR, Mulkerin D, Lubner S, Richter K, Winterle N, Wims MB, Dietrich L, Winkler JM, Volk M, Kim K, Holen KD. A phase II study of capecitabine and lapatinib in advanced refractory colorectal adenocarcinoma: A Wisconsin Oncology Network study. *J Gastrointest Oncol.* 2012; 3(2):90-6.

[94] Sartore-Bianchi A, Trusolino L, Martino C, Bencardino K, Lonardi S, Bergamo F, Zagonel V, Leone F, Depetris I, Martinelli E, Troiani T, Ciardiello F, Racca P, Bertotti A, Siravegna G, Torri V, Amatu A, Ghezzi S, Marrapese G, Palmeri L, Valtorta E, Cassingena A, Lauricella C, Vanzulli A, Regge D, Veronese S, Comoglio PM, Bardelli A, Marsoni S, Siena S. Dual-targeted therapy with

trastuzumab and lapatinib in treatment-refractory, KRAS codon 12/13 wild-type, HER2-positive metastatic colorectal cancer (HERACLES): a proof-of-concept, multicentre, open-label, phase 2 trial. *Lancet Oncol* 2016; 17: 738-746.

[95] Siena S, Bardelli A, Sartore-Bianchi A, Martino C, Siravegna G, Magri' A, Leone F, Zagonel V, Lonardi S, Amatu A, Tosi F, Racca P, Ponzetti A, Ciardiello F, Marsoni S. HER2 amplification as a 'molecular bait' for trastuzumab-emtansine (T-DM1) precision chemotherapy to overcome anti-HER2 resistance in HER2 positive metastatic colorectal cancer: the HERACLES-RESCUE trial. *J Clin Oncol* 2016; 34: 774.

[96] Trusolino L, Bertotti A, Lonardi S, Sartore-Bianchi A, Martino C, Cottino F, Vurchio V, Valtorta E, Lauricella C, Regge D, Vanzulli A, Zagonel V, Leone F, Racca P, Ciardiello F, Ardizzoni A, Marsoni S, Siena S. Pertuzumab and trastuzumab-emtansine in HER2-positive colorectal cancer: the HERACLES B trial. *Cancer Res* 2016; 76 (Suppl): Abstract CT082.

[97] Hainsworth JD, Meric-Bernstam F, Swanton C, Hurwitz H, Spigel DR, Sweeney C, Burris HA, Bose R, Guo S, Bernaards C, Beattie MS, Stein A, Brammer M, Kurzrock R. Targeted therapy for advanced solid tumors based on molecular profiles: Early results from MyPathway, an open-label, phase II umbrella basket study. *J Clin Oncol* 2016; 34 (suppl): 1151).

[98] Strickler JH, Niedzwiecki D, Zemla T, Cercek A, Fakih M, Kimmie Ng, Sanchez FA, Wu C, Peterson S, Bandel L, Grothey A, Bekaii-Saab TS. A phase II, open label study of tucatinib (ONT-380) combined with trastuzumab in patients with HER2+ metastatic colorectal cancer (mCRC)(MOUNTAINEER). *J Clin Oncol* 2017; 35 (suppl): TPS3624.

[99] Peterson S, de Vires P, Piasecki J, Rosler R. Tucatinib, a HER2 selective kinase inhibitor, is active in patients derived xenograft (PDX) models of HER2-amplified colorectal, esophageal and gastric cancers. *Ann Oncol* 2017; 28: v573-v594.

[100] ZW25 Effective in HER2-Positive Cancers. *Cancer Discov*. 2019 Jan; 9(1):8.

[101] Hickish T, Cassidy J, Propper D, Chau I, Falk S, Ford H, Iveson T, Braun M, Potter V, Macpherson IR, Finnigan H, Lee C, Jones H, Harrison M. A randomised, open-label phase II trial of afatinib versus cetuximab in patients with metastatic colorectal cancer. *Eur J Cancer*. 2014; 50(18):3136-44.

[102] Helene Senellart, Emmanuelle Samalin, Florence Castan, Christophe Borg, Antoine Adenis, Christelle De La Fouchardiere, David Malka, Veronique Guerin-Meyer, Eric Francois, Meher Ben Abdelghani, Eveline Boucher, Thierry Andre, Francois Ghiringhelli, Astrid Lievre, Trevor Stanbury, Jaafar Bennouna A multi-centric randomized phase II trial evaluating dual targeting of the EGFR with cetuximab and afatinib versus cetuximab alone in patients with chemotherapy refractory wtKRAS metastatic colorectal cancer (UCGI 25: A UNICANCER trial). Abstrac 3537 *J. of Clin Oncology* 34 no 15 supp May 2016).

[103] Yang M, Fang X, Li J, Xu D, Xiao Q, Yu S, Hu H, Weng S, Ding K, Yuan Y. Afatinib treatment for her-2 amplified metastatic colorectal cancer based on patient-derived xenograft models and next generation sequencing. *Cancer Biol Ther*. 2019; 20(4):391-396.

[104] Benvenuti, S, Sartore-Bianchi, A, Di Nicolantonio, F, Zanon, C, Moroni, M, Veronese, S, Siena, S, Bardelli, A, 2007. Oncogenic activation of the RAS/RAFsignaling pathway impairs the response of metastatic colorectal cancers toanti-epidermal growth factor receptor antibody therapies. *Cancer Res*. 67, 2643–2648.

[105] De Roock, W, Claes, B, Bernasconi, D, De Schutter, J, Biesmans, B, Fountzilas, G, Kalogeras, KT, Kotoula, V, Papamichael, D, Laurent-Puig, P, Penault-Llorca, F, Rougier, P, Vincenzi, B, Santini, D, Tonini, G, Cappuzzo, F, Frattini, M, Molinari, F, Saletti, P, De Dosso, S, Martini, M, Bardelli, A, Siena, S, Sartore-Bianchi, A, Tabernero, J, Macarulla, T, Di Fiore, F, Gangloff, AO, Ciardiello, F, Pfeiffer, P, Qvortrup, C, Hansen, TP, Van Cutsem, E, Piessevaux, H, Lambrechts, D, Delorenzi, M, Tejpar, S, 2010a. Effects of KRAS, BRAF, NRAS,

and PIK3CA mutations on the efficacy of cetuximab plus chemotherapy inchemotherapy-refractory metastatic colorectal cancer: a retrospectiveconsortium analysis. *Lancet Oncol.* 11, 753–762.

[106] De Roock, J, Jonker, DJ, Nicolantonio, F, Sartore-Bianchi, A, Tu, D, Siena, S, Lamba, S, Arena, S, Frattini, M, Piessevaux, H, Van Cutsem, E, O'Callaghan, CJ, Khambata-Ford, S, Zalcberg, JR, Simes, J, Karapetis, CS, Bardelli, A, Tejpar, S, 2010b. Association of KRAS p. G13D mutation with outcome inpatients with chemotherapy-refractory metastatic colorectal cancer treatedwith cetuximab. *JAMA* 304, 1812–1820.

[107] Zhao B, Wang L, Qiu H, Zhang M, Sun L, Peng P, Yu Q, Yuan X. Mechanisms of resistance to anti-EGFR therapy in colorectal cancer. *Oncotarget.* 2017; 17; 8(3):3980-4000.

[108] Bardelli A, Siena S. Molecular mechanisms of resistance to cetuximab and panitumumab in colorectal cancer. *J Clin Oncol.* 2010; 28:1254–61.

[109] Tejpar S, Celik I, Schlichting M, Sartorius U, Bokemeyer C, Van Cutsem E. Association of KRAS G13D tumor mutations with outcome in patients with metastatic colorectal cancer treated with first-line chemotherapy with or without cetuximab. *J Clin Oncol.* 2012; 30:3570–7.

[110] Gajate P, Sastre J, Bando I, Alonso T, Cillero L, Sanz J, Caldes T, Diaz-Rubio E. Influence of KRAS p.G13D mutation in patients with metastatic colorectal cancer treated with cetuximab. *Clin Colorectal Cancer.* 2012; 11:291–6.

[111] Peeters M, Douillard JY, Van Cutsem E, Siena S, Zhang K, Williams R, Wiezorek J. Mutant KRAS codon 12 and 13 alleles in patients with metastatic colorectal cancer: assessment as prognostic and predictive biomarkers of response to panitumumab. *J Clin Oncol.* 2013; 31:759–65.

[112] Douillard JY, Oliner KS, Siena S, Tabernero J, Burkes R, Barugel M, Humblet Y, Bodoky G, Cunningham D, Jassem J, Rivera F, Kocakova I, Ruff P, Błasińska-Morawiec M, Šmakal M, Canon JL, Rother M,

Williams R, Rong A, Wiezorek J, Sidhu R, Patterson SD. Panitumumab-FOLFOX4 treatment and RAS mutations in colorectal cancer. *N Engl J Med.* 2013; 369: 1023–34.

[113] Sartore-Bianchi, A, Bencardino, K, Di Nicolantonio, F, Pozzi, F, Funaioli, C, Gambi, V, Arena, S, Martini, M, Lamba, S, Cassingena, A, Schiavo, R, Bardelli, A, Siena, S. Integrated molecular dissection of the epidermal growth factorreceptor (EGFR) oncogenic pathway to predict response to EGFR-targetedmonoclonal antibodies in metastatic colorectal cancer. *Target Oncol.* 2010; 5, 19–28.

[114] Seymour MT, Brown SR, Middleton G, Maughan T, Richman S, Gwyther S, Lowe C, Seligmann JF, Wadsley J, Maisey N, Chau I, Hill M, Dawson L, Falk S, O'Callaghan A, Benstead K, Chambers P, Oliver A, Marshall H, Napp V, Quirke P.. Panitumumab and irinotecan versus irinotecan alone for patients with KRAS wild-type, fluorouracil-resistant advanced colorectal cancer (PICCOLO): a prospectively stratified randomised trial. *Lancet Oncol.* 2013; 14:749–59.

[115] van Geel RMJM, Tabernero J, Elez E, Bendell JC, Spreafico A, Schuler M, Yoshino T, Delord JP, Yamada Y, Lolkema MP, Faris JE, Eskens FALM, Sharma S, Yaeger R, Lenz HJ, Wainberg ZA, Avsar E, Chatterjee A, Jaeger S, Tan E, Maharry K, Demuth T, Schellens JHM. A Phase Ib Dose-Escalation Study of Encorafenib and Cetuximab with or without Alpelisib in Metastatic BRAF-Mutant *Colorectal Cancer Cancer Discov*. 2017; 7(6):610-619.

[116] Sartore-Bianchi, A, Moroni, M, Veronese, S, Carnaghi, C, Bajetta, E, Luppi, G, Sobrero, A, Barone, C, Cascinu, S, Colucci, G, Cortesi, E, Nichelatti, M, Gambacorta, M, Siena, S. Epidermal growth factor receptor gene copy number and clinical outcome of metastatic colorectal cancer treated withpanitumumab. *J. Clin. Oncol.* 2007; 25, 3238–3245.

[117] .Laurent-Puig P, Cayre A, Manceau G, Buc E, Bachet JB, Lecomte T, Rougier P, Lievre A, Landi B, Boige V, Ducreux M, Ychou M, Bibeau F, Bouché O, Reid J, Stone S, Penault-Llorca F. Analysis of PTEN, BRAF, and EGFR status in determining benefit from cetuximab

therapy in wild-type KRAS metastatic colon cancer. *J Clin Oncol.* 2009; 27(35):5924-30.

[118] Troiani, T, Martinelli, E, Napolitano, S, Morgillo, F, Belli, G, Cioffi, L, Ciardiello, F. Molecular aspects of resistance to biological and non-biological drugsand strategies to overcome resistance in colorectal cancer. *Curr. Med. Chem.* 2014; 21, 1639–1653.

[119] Zhao L, Vogt PK. Helical domain and kinase domain mutations in p110alpha of phosphatidylinositol 3-kinase induce gain of function by different mechanisms. *Proc Natl Acad Sci U S A*. 2008; 105:2652–7.

[120] Xu JM, Wang Y, Wang YL, Wang Y, Liu T, Ni M, Li MS, Lin L, Ge FJ, Gong C, Gu JY, Jia R, Wang HF, Chen YL, Liu RR, Zhao CH, Tan ZL, Jin Y, Zhu YP, Ogino S, Qian ZR. *PIK3CA* Mutations Contribute to Acquired Cetuximab Resistance in Patients with Metastatic Colorectal Cancer. *Clin Cancer Res.* 2017; 23(16):4602-4616.

[121] Mao C, Yang ZY, Hu XF, Chen Q, Tang JL. PIK3CA exon 20 mutations as a potential biomarker for resistance to anti-EGFR monoclonal antibodies in KRAS wild-type metastatic colorectal cancer: a systematic review and meta-analysis. *Ann Oncol.* 2012; 23:1518–25.

[122] Ma XT, Wang S, Ye YJ, Du RY, Cui ZR, Somsouk M. Constitutive activation of Stat3 signaling pathway in human colorectal carcinoma. *World J Gastroenterol*. 2004; 10:1569–73.

[123] Zhu QC, Gao RY, Wu W, Qin HL. Epithelial-mesenchymal transition and its role in the pathogenesis of colorectal cancer Asian *Pac J Cancer Prev.* 2013; 14(5):2689-98.

[124] Buck E, Eyzaguirre A, Barr S, Thompson S, Sennello R, Young D, Iwata KK, Gibson NW, Cagnoni P, and Haley JD. Loss of homotypic cell adhesion by epithelial-mesenchymal transition or mutation limits sensitivity to epidermal growth factor receptor inhibition. *Mol Cancer Ther.* 2007; 6: 532- 41.

[125] De Luca A, Arra C, D'Antonio A, Casamassimi A, Losito S, Ferraro P, Ciardiello F, Salomon DS, and Normanno N. Simultaneous blockage of different EGF-like growth factors results in efficient

growth inhibition of human colon carcinoma xenografts. *Oncogene.* 2000; 19: 5863-71.

[126] Gatenby, RA, Silva, AS, Gillies, RJ, Frieden, BR. Adaptive therapy. *Cancer Res*. 2009; 69, 4894–4903.

[127] Bouchahda M, Karaboue A, Saffroy R, Innominato P, Gorden L, Guettier C, Adam R, Levi F. Acquired KRAS mutations during progression of colorectal cancer metastases: possible implications for therapy and prognosis. *Cancer Chemother Pharmacol.* 2010; 66:605–9.

[128] Misale S, Arena S, Lamba S, Siravegna G, Lallo A, Hobor S, Russo M, Buscarino M, Lazzari L, Sartore-Bianchi A, Bencardino K, Amatu A, Lauricella C, Valtorta E, Siena S, Di Nicolantonio F, Bardelli A. Blockade of EGFR and MEK intercepts heterogeneous mechanisms of acquired resistance to anti-EGFR therapies in colorectal cancer. *Sci Transl Med.* 2014; 6:224ra26.

[129] Bertotti, A, Migliardi, G, Galimi, F, Sassi, F, Torti, D, Isella, C, Corà, D, DiNicolantonio, F, Buscarino, M, Petti, C, Ribero, D, Russolillo, N, Muratore, A, Massucco, P, Pisacane, A, Molinaro, L, Valtorta, E, Sartore-Bianchi, A, Risio, M, Capussotti, L, Gambacorta, M, Siena, S, Medico, E, Sapino, A, Marsoni, S, Comoglio, PM, Bardelli, A, Trusolino, L. A molecularly annotatedplatform of patient derived xenografts (xenopatients) identifies HER2 as aneffective therapeutic target in cetuximab-resistant colorectal, cancer. *Cancer Discov.* 2011; 1, 508–523.

[130] Yonesaka, K, Zejnullahu, K, Okamoto, I, Satoh, T, Cappuzzo, F, Souglakos, J, Ercan, D, Rogers, A, Roncalli, M, Takeda, M, Fujisaka, Y, Philips, J, Shimizu, T, Maenishi, O, Cho, Y, Sun, J, Destro, A, Taira, K, Takeda, K, Okabe, T, Swanson, J, Itoh, H, Takada, M, Lifshits, E, Okuno, K, Engelman, JA, Shivdasani, RA, Nishio, K, Fukuoka, M, Varella-Garcia, M, Nakagawa, K, Jänne, PA. Activation of ERBB2 signaling causes resistance to the EGFR directedtherapeutic antibody cetuximab. *Sci. Transl. Med.* 2011; 3, 99ra86.

[131] Reidy DL, Vakiani E, Fakih MG, Saif MW, Hecht JR, Goodman-Davis N, Hollywood E, Shia J, Schwartz J, Chandrawansa K, Dontabhaktuni A, Youssoufian H, Solit DB, Saltz LB. Randomized, phase II study of the insulin-like growth factor-1 receptor inhibitor IMC-A12, with or without cetuximab, in patients with cetuximab- or panitumumab-refractory metastatic colorectal cancer. *J Clin Oncol.* 2010; 28:4240–6.

[132] Sclafani F, Kim TY, Cunningham D, Kim TW, Tabernero J, Schmoll HJ, Roh JK, Kim SY, Park YS, Guren TK, Hawkes E, Clarke SJ, Ferry D, Frödin JE, Ayers M, Nebozhyn M, Peckitt C, Loboda A, Mauro DJ, Watkins DJ. A Randomized Phase II/III Study of Dalotuzumab in Combination With Cetuximab and Irinotecan in Chemorefractory, KRAS Wild-Type, Metastatic Colorectal Cancer. *J Natl Cancer Inst.* 2015; 107:djv258.

[133] Troiani T, Martinelli E, Napolitano S, Vitagliano D, Ciuffreda LP, Costantino S, Morgillo F, Capasso A, Sforza V, Nappi A, De Palma R, D'Aiuto E, Berrino L, Bianco R, Ciardiello F. Increased TGF-alpha as a mechanism of acquired resistance to the anti-EGFR inhibitor cetuximab through EGFR-MET interaction and activation of MET signaling in colon cancer cells. *Clin Cancer Res.* 2013; 19:6751–65.

[134] Liska D, Chen CT, Bachleitner-Hofmann T, Christensen JG, Weiser MR. HGF rescues colorectal cancer cells from EGFR inhibition via MET activation. *Clin Cancer Res.* 2011; 17:472–82.

[135] Krumbach R, Schuler J, Hofmann M, Giesemann T, Fiebig HH, Beckers T. Primary resistance to cetuximab in a panel of patient-derived tumour xenograft models: activation of MET as one mechanism for drug resistance. *Eur J Cancer.* 2011; 47: 1231–43.

[136] Bardelli A, Corso S, Bertotti A, Hobor S, Valtorta E, Siravegna G, Sartore-Bianchi A, Scala E, Cassingena A, Zecchin D, Apicella M, Migliardi G, Galimi F, Lauricella C, Zanon C, Perera T, Veronese S, Corti G, Amatu A, Gambacorta M, Diaz LA Jr, Sausen M, Velculescu VE, Comoglio P, Trusolino L, Di Nicolantonio F, Giordano S, Siena S.. Amplification of the MET receptor drives resistance to anti-EGFR therapies in colorectal cancer. *Cancer Discov.* 2013; 3:658–73.

[137] Montagut C, Dalmases A, Bellosillo B, Crespo M, Pairet S, Iglesias M, Salido M, Gallen M, Marsters S, Tsai SP, Minoche A, Seshagiri S, Serrano S, Himmelbauer H, Bellmunt J, Rovira A, Settleman J, Bosch F, Albanell J. Identification of a mutation in the extracellular domain of the Epidermal Growth Factor Receptor conferring cetuximab resistance in colorectal cancer. *Nat Med.* 2012; 18:221–3.

[138] Esposito C, Rachiglio AM, La Porta ML, Sacco A, Roma C, Iannaccone A, Tatangelo F, Forgione L, Pasquale R, Barbaro A, Botti G, Ciardiello F, Normanno N. The S492R EGFR ectodomain mutation is never detected in KRAS wild-type colorectal carcinoma before exposure to EGFR monoclonal antibodies. *Cancer Biol Ther.* 2013; 14:1143–6.

[139] Ruckhäberle, E, Rody, A, Engels, K, Gaetje, R, von Minckwitz, G, Schiffmann, S, Grösch, S, Geisslinger, G, Holtrich, U, Karn, T, Kaufmann, M. Micro array analysis of altered sphingolipid metabolism reveals prognostic significance ofsphingosine kinase 1 in breast cancer. *Breast Cancer Res. Treat.* 2008, 112, 41–52.

[140] Sukocheva, O, Wang, L, Verrier, E, Vadas, MA, Xia, P, Restoring endocrineresponse in breast cancer cells by inhibition of the sphingosine kinase-1signaling pathway. *Endocrinology* 2009; 150, 4484–4492.

[141] Rosa, R, Marciano, R, Malapelle, U, Formisano, L, Nappi, L, D'Amato, C, D'Amato, V, Damiano, V, Marfè, G, Del Vecchio, S, Zannetti, A, Greco, A, De Stefano, A, Carlomagno, C, Veneziani, BM, Troncone, G, De Placido, S, Bianco, R. Sphingosine kinase 1 overexpression contributes to cetuximab resistance inhuman colorectal cancer models. *Clin. Cancer Res.* 2013; 19, 138–147.

[142] *SEER Cancer Statistics Factsheets: Chronic Myeloid Leukemia.* National Cancer Institute; Bethesda, MD: http://seer.cancer.gov/statfacts/html/cmyl.html.

[143] Bartram CR, de Klein A, Hagemeijer A, van Agthoven T, Geurts van Kessel A, Bootsma D, Grosveld G, Ferguson-Smith MA, Davies T, Stone M, et al. Translocation of c-ab1 oncogene correlates with the

presence of a Philadelphia chromosome in chronic myelocytic leukaemia. *Nature.* 1983; 306:277– 280.
[144] Rowley JD. Letter: A new consistent chromosomal abnormality in chronic myelogenous leukaemia identified by quinacrine fluorescence and Giemsa staining. *Nature.* 1973; 243:290–293).
[145] Deininger MW, Goldman JM, Melo JV. The molecular biology of chronic myeloid leukemia. *Blood.* 2000; 96(10):3343.–3356.
[146] Steelman LS, Pohnert SC, Shelton JG, Franklin RA, Bertrand FE, McCubrey JA. JAK/STAT, Raf/MEK/ERK, PI3K/Akt and BCR-ABL in cell cycle progression and leukemogenesis. *Leukemia.* 2004; 18(2): 189–218.
[147] Druker BJ, Tamura S, Buchdunger E, Ohno S, Segal GM, Fanning S, Zimmermann J, Lydon NB. Effects of a selective inhibitor of the Abl tyrosine kinase on the growth of Bcr-Abl positive cells. *Nat Med.* 1996 May; 2(5):561-6.
[148] Deininger M, Buchdunger E, Druker BJ. The development of imatinib as a therapeutic agent for chronic myeloid leukemia. *Blood.* 2005 Apr 1; 105(7):2640-53.
[149] Schindler T, Bornmann W, Pellicena P, Miller WT, Clarkson B, Kuriyan J. Structural mechanism for STI-571 inhibition of abelson tyrosine kinase. *Science.* 2000; 289(5486):1938-42.
[150] Druker BJ, Talpaz M, Resta DJ, Peng B, Buchdunger E, Ford JM, Lydon NB, Kantarjian H, Capdeville R, Ohno-Jones S, Sawyers CL. Efficacy and safety of a specific inhibitor of the BCR-ABL tyrosine kinase in chronic myeloid leukemia. *N Engl J Med.* 2001; 344:1031–1037.
[151] Kantarjian H, Sawyers C, Hochhaus A, Guilhot F, Schiffer C, Gambacorti-Passerini C, Niederwieser D, Resta D, Capdeville R, Zoellner U, Talpaz M, Druker B, Goldman J, O'Brien SG, Russell N, Fischer T, Ottmann O, Cony-Makhoul P, Facon T, Stone R, Miller C, Tallman M, Brown R, Schuster M, Loughran T, Gratwohl A, Mandelli F, Saglio G, Lazzarino M, Russo D, Baccarani M, Morra E; International STI571 CML Study Group. Hematologic and

cytogenetic responses to imatinib mesylate in chronic myelogenous leukemia. *N Engl J Med.* 2002; 346:645–652.

[152] O'Brien SG, Guilhot F, Larson RA, athmann I, Baccarani M, Cervantes F, Cornelissen JJ, Fischer T, Hochhaus A, Hughes T, Lechner K, Nielsen JL, Rousselot P, Reiffers J, Saglio G, Shepherd J, Simonsson B, Gratwohl A, Goldman JM, Kantarjian H, Taylor K, Verhoef G, Bolton AE, Capdeville R, Druker BJ; IRIS Investigators. Imatinib compared with interferon and low-dose cytarabine for newly diagnosed chronic-phase chronic myeloid leukemia. *N Engl J Med.* Mar 13; 2003 348(11):994–1004.

[153] Druker BJ, Guilhot F, O'Brien SG, Gathmann I, Kantarjian H, Gattermann N, Deininger MW, Silver RT, Goldman JM, Stone RM, Cervantes F, Hochhaus A, Powell BL, Gabrilove JL, Rousselot P, Reiffers J, Cornelissen JJ, Hughes T, Agis H, Fischer T, Verhoef G, Shepherd J, Saglio G, Gratwohl A, Nielsen JL, Radich JP, Simonsson B, Taylor K, Baccarani M, So C, Letvak L, Larson RA; IRIS Investigators. Five-year follow-up of patients receiving imatinib for chronic myeloid leukemia. *N Engl J Med.;* 2006 355(23):2408–2417.

[154] Sawyers CL, Hochhaus A, Feldman E, Goldman JM, Miller CB, Ottmann OG, Schiffer CA, Talpaz M, Guilhot F, Deininger MW, Fischer T, O'Brien SG, Stone RM, Gambacorti-Passerini CB, Russell NH, Reiffers JJ, Shea TC, Chapuis B, Coutre S, Tura S, Morra E, Larson RA, Saven A, Peschel C, Gratwohl A, Mandelli F, Ben-Am M, Gathmann I, Capdeville R, Paquette RL, Druker BJ. Imatinib induces hematologic and cytogenetic responses in patients with chronic myelogenous leukemia in myeloid blast crisis: results of a phase II study. *Blood.* 2002 May 15; 99(10):3530-9.

[155] Shah NP. Loss of response to imatinib: mechanisms and management. *Hematology Am Soc Hematol Educ Program.* 2005: 183-7.

[156] Quintás-Cardama A, Cortes J. Molecular biology of bcr-abl1-positive chronic myeloid leukemia. *Blood.* 2009; 113(8):1619-30.

[157] Illmer T, Schaich M, Platzbecker U, Freiberg-Richter J, Oelschlägel U, von Bonin M, Pursche S, Bergemann T, Ehninger G, Schleyer E. P-glycoprotein-mediated drug efflux is a resistance mechanism of

chronic myelogenous leukemia cells to treatment with imatinib mesylate. *Leukemia.* 2004; 18(3):401-8.

[158] Hochhaus A, O'Brien SG, Guilhot F, ruker BJ, Branford S, Foroni L, Goldman JM, Müller MC, Radich JP, Rudoltz M, Mone M, Gathmann I, Hughes TP, Larson RA; IRIS Investigators. Six-year follow-up of patients receiving imatinib for the first-line treatment of chronic myeloid leukemia. *Leukemia.* 2009; 23(6):1054–1061.

[159] Kalmanti L, Saussele S, Lauseker M, Müller MC, Dietz CT, Heinrich L, Hanfstein B, Proetel U, Fabarius A, Krause SW, Rinaldetti S, Dengler J, Falge C, Oppliger-Leibundgut E, Burchert A, Neubauer A, Kanz L, Stegelmann F, Pfreundschuh M, Spiekermann K, Scheid C, Pfirrmann M, Hochhaus A, Hasford J, Hehlmann R. Safety and efficacy of imatinib in CML over a period of 10 years: data from the randomized CML-study IV. *Leukemia.* 2015; 29(5):1123–1132.

[160] Castagnetti F, Gugliotta G, Breccia M, Stagno F, Iurlo A, Albano F, Abruzzese E, Martino B, Levato L, Intermesoli T, Pregno P, Rossi G, Gherlinzoni F, Leoni P, Cavazzini F, Venturi C, Soverini S, Testoni N, Alimena G, Cavo M, Martinelli G, Pane F, Saglio G, Rosti G, Baccarani M; GIMEMA CML Working Party. Long-term outcome of chronic myeloid leukemia patients treated frontline with imatinib. *Leukemia.* 2015; 29(9):1823–1831.

[161] Zhang WW, Cortes JE, Yao H, hang L, Reddy NG, Jabbour E, Kantarjian HM, Jones D. Predictors of primary imatinib resistance in chronic myelogenous leukemia are distinct from those in secondary imatinib resistance. *J Clin Oncol.* 2009; 27:3642–3649.

[162] le Coutre P, Tassi E, Varella-Garcia M, Barni R, Mologni L, Cabrita G, Marchesi E, Supino R, Gambacorti-Passerini C. Induction of resistance to the Abelson inhibitor STI571 in human leukemic cells through gene amplification. *Blood.* 2000; 95:1758–1766.

[163] Gorre ME, Mohammed M, Ellwood K, Hsu N, Paquette R, Rao PN, Sawyers CL Clinical resistance to STI-571 cancer therapy caused by BCR-ABL gene mutation or amplification. *Science.* 2001; 293:876–880.

[164] Branford S, Rudzki Z, Walsh S, Grigg A, Arthur C, Taylor K, Herrmann R, Lynch KP, Hughes TP. High frequency of point mutations clustered within the adenosine triphosphate-binding region of BCR/ABL in patients with chronic myeloid leukemia or Ph-positive acute lymphoblastic leukemia who develop imatinib (STI571) resistance. *Blood.* 2002; 99(9):3472-5.

[165] Soverini S, Colarossi S, Gnani A, Rosti G, Castagnetti F, Poerio A, Iacobucci I, Amabile M, Abruzzese E, Orlandi E, Radaelli F, Ciccone F, Tiribelli M, di Lorenzo R, Caracciolo C, Izzo B, Pane F, Saglio G, Baccarani M, Martinelli G; GIMEMA Working Party on Chronic Myeloid Leukemia. Contribution of ABL kinase domain mutations to imatinib resistance in different subsets of Philadelphia-positive patients: by the GIMEMA Working Party on Chronic Myeloid Leukemia. *Clin Cancer Res.* 2006; 12:7374–7379.

[166] Branford S, Rudzki Z, Walsh S, Parkinson I, Grigg A, Szer J, Taylor K, Herrmann R, Seymour JF, Arthur C, Joske D, Lynch K, Hughes T. Detection of BCR-ABL mutations in patients with CML treated with imatinib is virtually always accompanied by clinical resistance, and mutations in the ATP phosphate-binding loop (P-loop) are associated with a poor prognosis. *Blood.* 2003 Jul 1; 102(1):276-83.

[167] Jabbour E, Kantarjian H, Jones D, O'Brien S, Luthra R, Garcia-Manero G, Francis Giles F, Mary Beth Rios MB Verstovsek S, Jorge Cortes J. Long-Term Incidence and Outcome of BCR-ABL Mutations in Patients (pts) with Chronic Myeloid Leukemia (CML) Treated with Imatinib Mesylate - P-Loop Mutations Are Not Associated with Worse Outcome. *Blood* (ASH Annual Meeting Abstracts) 2004; 104:1007.

[168] Kantarjian HM, Larson RA, Guilhot F, O'Brien SG, Mone M, Rudoltz M, Krahnke T, Cortes J, Druker BJ; International Randomized Study of Interferon and STI571 (IRIS) Investigators. Efficacy of imatinib dose escalation in patients with chronic myeloid leukemia in chronic phase. *Cancer.* 2009; 115:551–560.

[169] Jabbour E, Kantarjian HM, Jones D, han J, O'Brien S, Reddy N, Wierda WG, Faderl S, Garcia-Manero G, Verstovsek S, Rios MB,

Cortes J. Imatinib mesylate dose escalation is associated with durable responses in patients with chronic myeloid leukemia after cytogenetic failure on standard-dose imatinib therapy. *Blood.* 2009; 113:2154–2160.

[170] Cortes JE, Kantarjian HM, Goldberg SL, Powell BL, Giles FJ, Wetzler M, Akard L, Burke JM, Kerr R, Saleh M, Salvado A, McDougall K, Albitar M, Radich J. Rationale and Insight for Gleevec High-Dose Therapy (RIGHT) Trial Study Group. High-Dose imatinib in newly diagnosed chronicphase chronic myeloid leukemia: high rates of rapid cytogenetic and molecular responses. *J Clin Oncol.* 2009; 27:4754–4759.

[171] Baccarani M, Rosti G, Castagnetti F, Haznedaroglu I, Porkka K, Abruzzese E, Alimena G, Ehrencrona H, Hjorth-Hansen H, Kairisto V, Levato L, Martinelli G, Nagler A, Lanng Nielsen J, Ozbek U, Palandri F, Palmieri F, Pane F, Rege-Cambrin G, Russo D, Specchia G, Testoni N, Weiss-Bjerrum O, Saglio G, Simonsson B.. Comparison of imatinib 400 mg and 800 mg daily in the front-line treatment of high-risk, Philadelphia-positive chronic myeloid leukemia: a European Leukemia Net Study. *Blood.* 2009; 113:4497–4504.

[172] Jabbour E, Parikh SA, Kantarjian H, Cortes J. Chronic myeloid leukemia: mechanisms of resistance and treatment. *Hematol Oncol Clin North Am.*; 2011 25(5):981–995.

[173] Shah NP, Tran C, Lee FY, Chen P, Norris D, Sawyers CL. Overriding imatinib resistance with a novel ABL kinase inhibitor. *Science.* 2004; 305: 399–401.

[174] Weisberg E, Manley P, Mestan J, Cowan-Jacob S, Ray A, Griffin JD. AMN107 (nilotinib): a novel and selective inhibitor of BCR-ABL. *Br J Cancer.* 2006; 94:1765–9.

[175] Nicolini FE, Basak GW, Soverini S, Martinelli G, Mauro MJ, Müller MC, Hochhaus A, Chuah C, Dufva IH, Rege-Cambrin G, Saglio G, Michallet M, Labussière H, Morisset S, Hayette S, Etienne G, Olavarria E, Zhou W, Peter S, Apperley JF, Cortes J.. Allogeneic stem cell transplantation for patients harboring T315I BCR-ABL mutated leukemias *Blood,* 2011, 118 (20): 5697-570.

[176] Lasater, EA, Massi, ES, Stecula, A, Politi, J, Tan, SK, Smith, CC, Gunthorpe, M, Holmes, JP, Chehab, F, Sali, A. Novel TKI-resistant BCR-ABL1 gatekeeper residue mutations retain in vitro sensitivity to axitinib. *Leukemia.* 2016; 30, 1405-1409.

[177] Lombardo LJ, Lee FY, Chen P, Norris D, Barrish JC, Behnia K, Castaneda S, Cornelius LA, Das J, Doweyko AM, Fairchild C, Hunt JT, Inigo I, Johnston K, Kamath A, Kan D, Klei H, Marathe P, Pang S, Peterson R, Pitt S, Schieven GL, Schmidt RJ, Tokarski J, Wen ML, Wityak J, Borzilleri RM. Discovery of N-(2-chloro-6-methyl-phenyl)-2-(6-(4-(2hydroxyethyl)-piperazin-1-yl)-2-methylpyrimidin-4-ylamino)thiazole-5-carboxamide (BMS-354825), a dual Src/Abl kinase inhibitor with potent antitumor activity in preclinical assays. *J Med Chem.* 2004; 47:6658–6661.

[178] Schittenhelm MM, Shiraga S, Schroeder A, Corbin AS, Griffith D, Lee FY, Bokemeyer C, Deininger MW, Druker BJ, Heinrich MC. Dasatinib (BMS-354825), a dual SRC/ABL kinase inhibitor, inhibits the kinase activity of wild-type, juxtamembrane, and activation loop mutant KIT isoforms associated with human malignancies. *Cancer Res.* 2006; 66(1):473-81.

[179] Eck MJ, Manley PW. The interplay of structural information and functional studies in kinase drug design: insights from BCR-Abl. *Curr Opin Cell Biol.* 2009; 21(2):288-95.

[180] Cortes JE, Saglio G, Kantarjian HM, Baccarani M, Mayer J, Boqué C, Shah NP, Chuah C, Casanova L, Bradley-Garelik B, Manos G, Hochhaus A.. Final 5-year study results of DASISION: the Dasatinib Versus Imatinib Study in Treatment-Naïve Chronic Myeloid Leukemia Patients Trial. *J Clin Oncol* 2016; 34 (20): 2333-2340.

[181] Shah NP, Rousselot P, Schiffer CA, et Rea D, Cortes JE, Milone J, Mohamed H, Healey D, Kantarjian H, Hochhaus A, Saglio G. Dasatinib in imatinib-resistant or -intolerant chronic-phase, chronic myeloid leukemia patients: 7-year follow-up of study CA180-034. *Am J Hematol.* 2016; 91(9):869-874.

[182] Kantarjian H, Shah NP, Hochhaus A, Cortes J, Shah S, Ayala M, Moiraghi B, Shen Z, Mayer J, Pasquini R, Nakamae H, Huguet F,

Boqué C, Chuah C, Bleickardt E, Bradley-Garelik MB, Zhu C, Szatrowski T, Shapiro D, Baccarani M.. Dasatinib versus imatinib in newly diagnosed chronic-phase chronic myeloid leukemia. *N Engl J Med.* 2010; 362(24):2260-2270.

[183] Nakamae H, Fujisawa S, Ogura M, Uchida T, Onishi Y, Taniwaki M, Utsunomiya A, Matsue K, Takamatsu Y, Usuki K, Tanimoto M, Ishida Y, Ohashi K, Li L, Miyoshi M Dasatinib versus imatinib in Japanese patients with newly diagnosed chronic phase chronic myeloid leukemia: a subanalysis of the DASISION 5-year final report. *Int J Hematol.* 2017; 105(6):792-804.

[184] Shah NP, Kantarjian HM, Kim DW, Réa D, Dorlhiac-Llacer PE, Milone JH, Vela-Ojeda J, Silver RT, Khoury HJ, Charbonnier A, Khoroshko N, Paquette RL, Deininger M, Collins RH, Otero I, Hughes T, Bleickardt E, Strauss L, Francis S, Hochhaus A. Intermittent target inhibition with dasatinib 100 mg once daily preserves efficacy and improves tolerability in imatinib-resistant and -intolerant chronic-phase chronic myeloid leukemia. *J Clin Oncol.* 2008; 26(19):3204-3212.

[185] Soverini S, De Benedittis C, Castagnetti F, Gugliotta G, Mancini M, Bavaro L, Machova Polakova K, Linhartova J, Iurlo A, Russo D, Pane F, Saglio G, Rosti G, Cavo M, Baccarani M, Martinelli G. In chronic myeloid leukemia patients on second-line tyrosine kinase inhibitor therapy, deep sequencing of BCR-ABL1 at the time of warning may allow sensitive detection of emerging drug-resistant mutants. *BMC Cancer.* 2016 2; 16:572.

[186] Talpaz M, Shah NP, Kantarjian H, Donato N, Nicoll J, Paquette R, Cortes J, O'Brien S, Nicaise C, Bleickardt E, Blackwood-Chirchir MA, Iyer V, Chen TT, Huang F, Decillis AP, Sawyers CL Dasatinib in imatinib-resistant Philadelphia chromosome-positive leukemias. *N Engl J Med.* 2006; 354(24):2531-2541.

[187] Quintas-Cardama A, Kantarjian H, O'Brien S, Borthakur G, Bruzzi J, Munden R, Cortes J. Pleural effusion in patients with chronic myelogenous leukemia treated with dasatinib after imatinib failure. *J Clin Oncology* 2007: 25: 3908-3914.

[188] Wang X, Roy A, Hochhaus A, Kantarjian HM, Chen T-T, Shah NP. Differential effects of dosing regimen on the safety and efficacy of dasatinib: retrospective exposure–response analysis of a Phase III study. *Clin Pharmacol.* 2013; 5: 85-97.

[189] de Lavallade H, Punnialingam S, Milojkovic D, Bua M, Khorashad JS, Gabriel IH, Chaidos A, Olavarria E, Goldman JM, Apperley JF, Marin D. Pleural effusions in patients with chronic myeloid leukaemia treated with dasatinib may have an immune-mediated pathogenesis. *Br J Haematol.* 2008; 141(5):745-747.

[190] Bergeron A, Réa D, Levy V, Picard C, Meignin V, Tamburini J, Bruzzoni-Giovanelli H, Calvo F, Tazi A, Rousselot P. Lung abnormalities after dasatinib treatment for chronic myeloid leukemia: a case series. *Am J Respir Crit Care Med.* 2007; 176(8):814-8.

[191] Baumgart B, Guha M, Hennan J, Li J, Woicke J, Simic D, Graziano M, Wallis N, Sanderson T, Bunch RT In vitro and in vivo evaluation of dasatinib and imatinib on physiological parameters of pulmonary arterial hypertension. *Cancer Chemother Pharmacol.* 2017; 79(4):711-723.

[192] *Tasigna (nilotinib) [package insert]*. Novartis Pharmaceuticals Corporation: East Hanover, NJ, USA, 2015.

[193] Saglio G, Kim DW, Issaragrisil S, le Coutre P, Etienne G, Lobo C, Pasquini R, Clark RE, Hochhaus A, Hughes TP, Gallagher N, Hoenekopp A, Dong M, Haque A, Larson RA, Kantarjian HM; ENESTnd Investigators. Nilotinib versus imatinib for newly diagnosed chronic myeloid leukemia. *N Engl J Med* 2010; 362: 2251–2259.

[194] Manley PW, Cowan-Jacob SW, Fendrich G, Metan J. Molecular interactions between the highly selective pan-Bcr-Abl inhibitor, AMN107, and the tyrosine kinase domain of Abl. *Blood.* 2005; 106:940.

[195] White DL, Saunders VA, Dang P, Engler J, Zannettino ACW, Cambareri A, Quinn S, Manley P, Hughes TP. Oct-1 mediated influx is a key determinant of the intracellular uptake of imatinib but not

AMN107; reduced OCT-1 activity is the cause of low *in vitro* sensitivity to imatinib. *Blood* 2006; 108(2):697-704.

[196] Kantarjian HM, Cortes J, Guilhot F, Hochhaus A, Baccarani M, Lokey L. Diagnosis and management of chronic myeloid leukemia: a survey of American and European practice patterns. *Cancer.* 2007 Apr 1; 109(7):1365-75.

[197] Hughes TP, Saglio G, Kantarjian HM, Guilhot F, Niederwieser D, Rosti G, Nakaseko C, De Souza CA, Kalaycio ME, Meier S, Fan X, Menssen HD, Larson RA, Hochhaus A. Early molecular response predicts outcomes in patients with chronic myeloid leukemia in chronic phase treated with frontline nilotinib or imatinib. *Blood.* 2014; 123: 1353–1360.

[198] Hochhaus A, Saglio G, Larson RA, Kim DW, Etienne G, Rosti G, De Souza C, Kurokawa M, Kalaycio ME, Hoenekopp A, Fan X, Shou Y, Kantarjian HM, Hughes TP. Nilotinib is associated with a reduced incidence of BCR-ABL mutations versus imatinib in patients with newly diagnosed chronic myeloid leukemia in chronic phase. *Blood* 2013; 121: 3703–3708.

[199] Baccarani M, Cortes J, Pane F, Niederwieser D, Saglio G, Apperley J, Cervantes F, Deininger M, Gratwohl A, Guilhot F, Hochhaus A, Horowitz M, Hughes T, Kantarjian H, Larson R, Radich J, Simonsson B, Silver RT, Goldman J, Hehlmann R; European Leukemia Net. Chronic myeloid leukemia: An update of concepts and management recommendations of European LeukemiaNet. *J Clin Oncol.* 2009; 27:6041–6051.

[200] Cortes JE, Kim DW, Kantarjian HM, Brümmendorf TH, Dyagil I, Griskevicius L, Malhotra H, Powell C, Gogat K, Countouriotis AM, Gambacorti-Passerini C Bosutinib versus imatinib in newly diagnosed chronic-phase chronic myeloid leukemia: results from the BELA trial. *J Clin Oncol.* 2012; 30(28):3486-92.

[201] Hochhaus A, Kantarjian HM, Baccarani M, Lipton JH, Apperley JF, Druker BJ, Facon T, Goldberg SL, Cervantes F, Niederwieser D, Silver RT, Stone RM, Hughes TP, Muller MC, Ezzeddine R, Countouriotis AM, Shah NP. Dasatinib induces notable hematologic

and cytogenetic responses in chronic-phase chronic myeloid leukemia after failure of imatinib therapy. *Blood.* 2007; 109:2303–2309.

[202] Bristol-Myers Squibb. *Sprycel (Dasatinib) tablet for oral use [package insert] Princeton*, NJ: Bristol-Myers Squibb; 2009.

[203] Novartis. *Gleevec (Imatinib Mesylate) tablets for oral use [package insert]* East Hanover, NJ: Novartis; 2009.

[204] Bartolovic K, Balabanov S, Hartmann U, Komor M, Boehmler AM, Bühring HJ, Möhle R, Hoelzer D, Kanz L, Hofmann WK, Brümmendorf TH. Inhibitory effect of imatinib on normal progenitor cells in vitro. *Blood.* 2004; 103:523–529.

[205] Cortes JE, Kantarjian H, Shah NP, Bixby D, Mauro MJ, Flinn I, O'Hare T, Hu S, Narasimhan NI, Rivera VM, Clackson T, Turner CD, Haluska FG, Druker BJ, Deininger MW, Talpaz M. Ponatinib in refractory Philadelphia chromosome-positive leukemias. *N Engl J Med.* 2012; 367(22):2075-88. doi: 10.1056/NEJMoa1205127.

[206] O'Hare T, Shakespeare WC, Zhu X, Eide CA, Rivera VM, Wang F, Adrian LT, Zhou T, Huang WS, Xu Q, Metcalf CA 3rd, Tyner JW, Loriaux MM, Corbin AS, Wardwell S, Ning Y, Keats JA, Wang Y, Sundaramoorthi R, Thomas M, Zhou D, Snodgrass J, Commodore L, Sawyer TK, Dalgarno DC, Deininger MW, Druker BJ, Clackson T. AP24534, a pan-BCR-ABL inhibitor for chronic myeloid leukemia, potently inhibits the T315I mutant and overcomes mutation-based resistance. *Cancer Cell.* 2009; 16(5):401-12. doi: 10.1016/j.ccr.2009.09.028.

[207] Baccarani M, Deininger MW, Rosti G, ochhaus A, Soverini S, Apperley JF, Cervantes F, Clark RE, Cortes JE, Guilhot F, Hjorth-Hansen H, Hughes TP, Kantarjian HM, Kim DW, Larson RA, Lipton JH, Mahon FX, Martinelli G, Mayer J, Müller MC, Niederwieser D, Pane F, Radich JP, Rousselot P, Saglio G, Saußele S, Schiffer C, Silver R, Simonsson B, Steegmann JL, Goldman JM, Hehlmann R. European LeukemiaNet recommendations for the management of chronic myeloid leukemia: 2013. *Blood.* 2013; 122(6):872-884.

[208] *Iclusig [package insert]*. Cambridge, MA: ARIAD Pharmaceuticals, Inc, a wholly owned subsidiary of Takeda Pharmaceutical Company Limited; 2016.

[209] *-Iclusig [summary of product characteristics]*. Leatherhead, United Kingdom: ARIAD Pharmaceuticals, Inc, a wholly owned subsidiary of Takeda Pharmaceutical Company Limited; 2017.

[210] Cortes JE, Kim DW, Pinilla-Ibarz J, e Coutre P, Paquette R, Chuah C, Nicolini FE, Apperley JF, Khoury HJ, Talpaz M, DiPersio J, DeAngelo DJ, Abruzzese E, Rea D, Baccarani M, Müller MC, Gambacorti-Passerini C, Wong S, Lustgarten S, Rivera VM, Clackson T, Turner CD, Haluska FG, Guilhot F, Deininger MW, Hochhaus A, Hughes T, Goldman JM, Shah NP, Kantarjian H; PACE Investigators A phase 2 trial of ponatinib in Philadelphia chromosome-positive leukemias. *N Engl J Med.* 2013; 369(19):1783-1796.

[211] Shah NP, Kim DW, Kantarjian H, Rousselot P, Llacer PE, Enrico A, Vela-Ojeda J, Silver RT, Khoury HJ, Müller MC, Lambert A, Matloub Y, Hochhaus. Potent, transient inhibition of BCR-ABL with dasatinib 100 mg daily achieves rapid and durable cytogenetic responses and high transformation-free survival ratesin chronic phase chronic myeloid leukemia patients with resistance, suboptimal response or intolerance to imatinib. *Haematologica*. 2010; 95(2):232-240.

[212] *Tasigna [package insert]*. East Hanover, NJ: Novartis Pharmaceuticals Corporation; 2017.

[213] Khoury HJ, Cortes JE, Kantarjian HM, Gambacorti-Passerini C, Baccarani M, Kim DW, Zaritskey A, Countouriotis A, Besson N, Leip E, Kelly V, Brümmendorf TH. Bosutinib is active in chronic phase chronic myeloid leukemia after imatinib and dasatinib and/or nilotinib therapy failure. *Blood*. 2012; 119(15):3403-3412.

[214] Nicolini FE, Ibrahim AR, Soverini S, Martinelli G, Müller MC, Hochhaus A, Dufva IH, Kim DW, Cortes J, Mauro MJ, Chuah C, Labussière H, Morisset S, Roche-Lestienne C, Lippert E, Hayette S, Peter S, Zhou W, Maguer-Satta V, Michallet M, Goldman J, Apperley JF, Mahon FX, Marin D, Etienne G. The BCR-ABLT315I mutation compromises survival in chronic phase chronic myelogenous

leukemia patients resistant to tyrosine kinase inhibitors, in a matched pair analysis. *Haematologica.* 2013; 98(10):1510-1516.

[215] Mauro MJ, Cortes JE, Kim DW, Pinilla-Ibarz J, le Coutre P, Paquette R, Chuah C, Nicolini FE, Apperley Khoury HJ, Talpaz M, DiPersio JM, DeAngelo DJ, Abruzzese E, Rea D, Baccarani M, Muller MC, Gambacorti-Passerini C, Wong S, Dorer DJ, Knickerbocker DK, Rivera VM, Clackson T, Turner CD, Haluska FG, Guilhot F, Deininger MW, Hochhaus A, Hughes T, Goldman JM, Shah N Kantarjian HM and The PACE Study Group Multivariate analyses of the clinical and molecular parameters associated with efficacy and safety in patients with chronic myeloid leukemia (CML) and Philadelphia chromosome-positive acute lymphoblastic leukemia (Ph1 ALL) treated with ponatinib in the PACE trial [abstract]. *Blood.* 2012; 120(21). Abstract 3747.

[216] Deininger MW, Hodgson JG, Shah NP, Cortes JE, Kim DW, Nicolini FE, Talpaz M, Baccarani M, Müller MC, Li J, Parker WT, Lustgarten S, Clackson T, Haluska FG, Guilhot F, Kantarjian HM, Soverini S, Hochhaus A, Hughes TP, Rivera VM, Branford S. Compound mutations in BCR-ABL1 are not major drivers of primary or secondary resistance to ponatinib in CP-CML patients. *Blood.* 2016; 127(6):703-712.

[217] Steegmann JL, Baccarani M, Breccia M, Casado LF, García-Gutiérrez V, Hochhaus A, Kim DW, Kim TD, Khoury HJ, Le Coutre P, Mayer J, Milojkovic D, Porkka K, Rea D, Rosti G, Saussele S, Hehlmann R, Clark RE European Leukemia Net recommendations for the management and avoidance of adverse events of treatment in chronic myeloid leukaemia. *Leukemia.* 2016; 30(8):1648-1671.

[218] Moslehi JJ. Cardiovascular toxic effects of targeted cancer therapies. *N Engl J Med.* 2016; 375(15):1457-1467.

[219] Stemhagen A, Adib D, Lustgarten S, McGarry L, Du Moulin R, Santillana S. The OMNI patient registry: a prospective observational registry to assess vascular safety in patients with CML and Ph+ ALL treated with ponatinib [abstract]. *J Clin Oncol.* 2017; 35(suppl 15). Abstract TPS7073.

[220] Loren CP, Aslan JE, Rigg RA, Nowak MS, Healy LD, Gruber A, Druker BJ, McCarty OJ. The BCRABL inhibitor ponatinib inhibits platelet immunoreceptor tyrosine-based activation motif (ITAM) signaling, platelet activation and aggregate formation under shear. *Thromb Res.* 2015; 135(1):155-160.

[221] Rea D, Mirault T, Cluzeau T, Gautier JF, Guilhot F, Dombret H, Messas E. Early onset hypercholesterolemia induced by the 2nd-generation tyrosine kinase inhibitor nilotinib in patients with chronic phase-chronic myeloid leukemia. *Haematologica.* 2014; 99(7):1197-1203.

[222] Hochhaus A, Pinilla-Ibarz J, Kim DW, Pinilla-Ibarz J, le Coutre P, Paquette R, Chuah C, Nicolini FE, Apperley Khoury HJ, Talpaz M, Baccarani M, Lustgarten S, Rivera MV, Haluska FG, Guilhot F, Deininger MW. Hughes T, Shah N. Kantarjian HM, Talpaz M, Cortes J. Clinical impact of dose modification and dose intensity on response to ponatinib in patients with Philadelphia chromosome-positive leukemias. *Poster presented at the annual meeting of the American Society of Clinical Oncology*; 30 May to 3 June 2014; Chicago, IL. Poster 7084.

[223] Dorer DJ, Knickerbocker RK, Baccarani M, Cortes JE, Hochhaus A, Talpaz M, Haluska FG. Impact of dose intensity of ponatinib on selected adverse events: multivariate analyses from a pooled population of clinical trial patients. *Leuk Res.* 2016; 48:84-91.

[224] DiNitto JP, Wu JC. Molecular mechanisms of drug resistance in tyrosine kinases cAbl and cKit. *Crit Rev Biochem Mol Biol.* 2011 Aug; 46(4):295-309.

[225] von Bubnoff N, Schneller F, Peschel C, Duyster J. BCR–ABL gene mutations in relation to clinical resistance of Philadelphia chromosome-positive leukaemia to STI571: a prospective study. *Lancet* 2002; 359: 487–491.

[226] von Bubnoff N, Schneller F, Peschel C, Duyster J. Different Bcr–Abl gene mutations can cause clinical resistance of Philadelphia-positive leukemia towards STI571. *Blood* 2001; 98: 770a (Abstract).

[227] Druker BJ, Mauro MJ. STI571: targeting BCR–ABL as therapy for CML. *Oncologist* 2001; 6: 233–238.

[228] Corbin AS, Buchdunger E, Furet P, Druker BJ. Analysis of the structural basis of specifity of inhibition of the Abl kinase by STI571. *J Biol Chem* 2002; 277: 32214–32219.

[229] Vajpai N, Strauss A, Fendrich G, Cowan-Jacob SW, Manley PW, Grzesiek S, Jahnke W. Solution conformations and dynamics of ABL kinase-inhibitor complexes determined by NMR substantiate the different binding modes of imatinib/nilotinib and dasatinib. *J Biol Chem.* 2008; 283:18292–302.

[230] Jabbour E, Hochhaus A, Cortes J, La Rosée P, Kantarjian HM. Choosing the best treatment strategy for chronic myeloid leukemia patients resistant to imatinib: weighing the efficacy and safety of individual drugs with BCR-ABL mutations and patient history. *Leukemia.* 2010; 24:6.

[231] White DL, Saunders VA, Dang P, Engler J, Zannettino AC, Cambareri AC, Quinn SR, Manley PW, Hughes TP. OCT-1-mediated influx is a key determinant of the intracellular uptake of imatinib but not nilotinib (AMN107): reduced OCT-1 activity is the cause of low in vitro sensitivity to imatinib. *Blood.* 2006; 108:697–704.

[232] Soverini S, Branford S, Nicolini FE, Talpaz M, Deininger MW, Martinelli G, Muller MC, Radich JP, Shah NP. Implications of BCR-ABL1 kinase domain-mediated resistance in chronic myeloid leukemia. *Leuk Res.* 2014; 38:10–20.

[233] Zabriskie MS, Eide CA, Tantravahi SK, Vellore NA, Estrada J, Nicolini FE, Khoury HJ, Larson RA, Konopleva M, Cortes JE, Kantarjian H, Jabbour EJ, Kornblau SM, Lipton JH, Rea D, Stenke L, Barbany G, Lange T, Hernández-Boluda JC, Ossenkoppele GJ, Press RD, Chuah C, Goldberg SL, Wetzler M, Mahon FX, Etienne G, Baccarani M, Soverini S, Rosti G, Rousselot P, Friedman R, Deininger M, Reynolds KR, Heaton WL, Eiring AM, Pomicter AD, Khorashad JS, Kelley TW, Baron R, Druker BJ, Deininger MW, O'Hare T.. BCR-ABL1 compound mutations combining key kinase

domain positions confer clinical resistance to ponatinib in Ph chromosome-positive leukemia. *Cancer Cell.* 2014; 26:428–42.

[234] Gibbons DL, Pricl S, Posocco P, Laurini E, Fermeglia M, Sun H, Talpaz M, Donato N, Quintas-Cardama A. Molecular dynamics reveal BCR-ABL1 polymutants as a unique mechanism of resistance to PAN-BCR-ABL1 kinase inhibitor therapy. *Proc Natl Acad Sci U S A.* 2014; 111:3550.

[235] Soverini S, De Benedittis C, Castagnetti F, Gugliotta G, Mancini M, Bavaro L, Machova Polakova K, Linhartova J, Iurlo A, Russo D, Pane F, Saglio G, Rosti G, Cavo M, Baccarani M, Martinelli G. In chronic myeloid leukemia patients on second-line tyrosine kinase inhibitor therapy, deep sequencing of BCR-ABL1 at the time of warning may allow sensitive detection of emerging drug-resistant mutants. *BMC Cancer.* 2016; 16: 572.

[236] Redaelli S, Piazza R, Rostagno R, Magistroni V, Perini P, Marega M, Gambacorti-Passerini C, Boschelli F..Activity of bosutinib, dasatinib, and nilotinib against 18 imatinib-resistant BCR/ABL mutants. *J Clin Oncol.* 2009; 27:469–71.

[237] Cortes JE, Kim DW, Pinilla-Ibarz J, le Coutre PD, Paquette R, Chuah C, Nicolini FE, Apperley JF, Khoury HJ, Talpaz M, DeAngelo DJ, Abruzzese E, Rea D, Baccarani M, Müller MC, Gambacorti-Passerini C, Lustgarten S, Rivera VM, Haluska FG, Guilhot F, Deininger MW, Hochhaus A, Hughes TP, Shah NP, Kantarjian HM Ponatinib efficacy and safety in Philadelphia chromosome-positive leukemia: final 5-year results of the phase 2 PACE trial. *Blood.* 2018; 132(4):393-404.

[238] Burgess MR, Skaggs BJ, Shah NP, Lee FY, Sawyers CL. Comparative analysis of two clinically active BCR-ABL kinase inhibitors reveals the role of conformation-specific binding in resistance. *Proc Natl Acad Sci U S A.* 2005; 102:3395–3400.

[239] White DL, Dang P, Engler J, Frede A, Zrim S, Osborn M, Saunders VA, Manley PW, Hughes TP. Functional activity of the OCT-1 protein is predictive of long-term outcome in patients with chronic-phase chronic myeloid leukemia treated with imatinib. *J Clin Oncol.* 2010; 28:2761–7.

[240] White DL, Saunders VA, Dang P, Engler J, Venables A, Zrim S, Zannettino A, Lynch K, Manley PW, Hughes T. Most CML patients who have a suboptimal response to imatinib have low OCT-1 activity: higher doses of imatinib may overcome the negative impact of low OCT-1 activity. *Blood.* 2007; 110:4064–72.

[241] Lu L, Saunders VA, Leclercq TM, Hughes TP, White DL Ponatinib is not transported by ABCB1, ABCG2 or OCT-1 in CML cells. *Leukemia.* 2015; 29:1792–4.

[242] Eiring AM, Kraft IL, BD Page BD, 'Hare TO, PT GunningPT and MW Deininger MW. STAT3 as a mediator of BCR-ABL1-independent resistance in chronic myeloid leukemia. *Leuk Suppl.* 2014; 3(Suppl 1): S5-6.

[243] Donato NJ, Wu JY, Stapley J, Gallick G, Lin H, Arlinghaus R, Talpaz M. BCR-ABL independence and LYN kinase overexpression in chronic myelogenous leukemia cells selected for resistance to STI571. *Blood.* 2003 Jan 15; 101(2):690-8.

[244] Wu J, Meng F, Lu H, Kong L, Bornmann W, Peng Z, Talpaz M, Donato NJ. Lyn regulates BCR-ABL and Gab2 tyrosine phosphorylation and c-Cbl protein stability in imatinib-resistant chronic myelogenous leukemia cells. *Blood.* 2008; 111(7):3821-9.

[245] Ma L, Shan Y, Bai R, Xue L, Eide CA, Ou J, Zhu LJ, Hutchinson L, Cerny J, Khoury HJ, Sheng Z, Druker BJ, Li S, Green MR A therapeutically targetable mechanism of BCR-ABL-independent imatinib resistance in chronic myeloid leukemia. *Sci Transl Med.* 2014 Sep 3; 6(252):252ra121.

[246] Sharma SV, Gajowniczek P, Way IP, Lee DY, Jiang J, Yuza Y, Classon M, Haber DA, Settleman J. A common signaling cascade may underlie "addiction" to the Src, BCR-ABL, and EGF receptor oncogenes. Cancer Cell. 2006 Nov; 10(5):425-35. *Erratum in: Cancer Cell.* 2009; 16(5):448.

[247] Logue JS, Morrison DK. Complexity in the signaling network: insights from the use of targeted inhibitors in cancer therapy. *Genes Dev*. 2012 Apr 1; 26(7):641-50.

[248] Duy C, Hurtz C, Shojaee S, Cerchietti L, Geng H, Swaminathan S, Klemm L, Kweon SM, Nahar R, Braig M, Park E, Kim YM, Hofmann WK, Herzog S, Jumaa H, Koeffler HP, Yu JJ, Heisterkamp N, Graeber TG, Wu H, Ye BH, Melnick A, Müschen M.BCL6 enables Ph+ acute lymphoblastic leukaemia cells to survive BCR-ABL1 kinase inhibition. *Nature.* 2011 May 19; 473(7347):384-8.

[249] Graham SM, Jorgensen HG, Allan E, Pearson C, Alcorn MJ, Richmond L, Holyoake TL.. Primitive, quiescent, Philadelphia-positive stem cells from patients with chronic myeloid leukemia are insensitive to STI571 in vitro. *Blood* 2002; 99: 319 –325.

[250] Corbin AS, Agarwal A, Loriaux M, Cortes J, Deininger MW, Druker BJ. Human chronic myeloid leukemia stem cells are insensitive to imatinib despite inhibition of BCR-ABL activity. *J Clin Invest* 2011; 121: 396 –409.

[251] Burchert A, Wang Y, Cai D, von Bubnoff N, Paschka P, Müller-Brüsselbach S, Ottmann OG, Duyster J, Hochhaus A, Neubauer A. Compensatory PI3-kinase/Akt/mTor activation regulates imatinib resistance development. *Leukemia.* 2005; 19:1774–82.

[252] Okabe S, Tauchi T, Tanaka Y, Kitahara T, Kimura S, Maekawa T, Ohyashiki K. Efficacy of the dual PI3K and mTOR inhibitor NVP-BEZ235 in combination with nilotinib against BCR-ABL-positive leukemia cells involves the ABL kinase domain mutation. *Cancer Biol Ther.* 2014; 15:207–15.

[253] Wagle M, Eiring AM, Wongchenko M, Lu S, Guan Y, Wang Y, Lackner M, Amler L, Hampton G, Deininger MW, O'Hare T, Yan Y. A role for FOXO1 in BCR-ABL1-independent tyrosine kinase inhibitor resistance in chronic myeloid leukemia. *Leukemia.* 2016; 30:1493–501. 54.

[254] Chu S, Holtz M, Gupta M, Bhatia R. BCR/ABL kinase inhibition by imatinib mesylate enhances MAP kinase activity in chronic myelogenous leukemia CD34+ cells. *Blood.* 2004; 103:3167–74.

[255] Packer LM, Rana S, Hayward R, O'Hare T, Eide CA, Rebocho A, Heidorn S, Zabriskie MS, Niculescu-Duvaz I, Druker BJ, Springer C, Marais R. Nilotinib and MEK inhibitors induce synthetic lethality

through paradoxical activation of RAF in drug-resistant chronic myeloid leukemia. *Cancer Cell.* 2011; 13;20(6):715-27.

[256] Suzuki M, Abe A, Imagama S, Nomura Y, Tanizaki R, Minami Y, Hayakawa F, Ito Y, Katsumi A, Yamamoto K, Emi N, Kiyoi H, Naoe TBCR-ABL-independent and RAS/MAPK pathway-dependent form of imatinib resistance in Ph-positive acute lymphoblastic leukemia cell line with activation of EphB4. *Eur J Haematol.* 2010; 84:229–38.

[257] Marfe G. Di Stefano C. Drug Resistance in Chronic Myelogenous Leukemia Caused by Sphingosine Kinases *J Blood Disord.* 2014; 1(2): 1010.

[258] Salas A, Ponnusamy S, Senkal CE, Meyers-Needham M, Selvam SP, Saddoughi SA, Apohan E, Sentelle RD, Smith C, Gault CR, Obeid LM, El-Shewy HM, Oaks J, Santhanam R, Marcucci G, Baran Y, Mahajan S, Fernandes D, Stuart R, Perrotti D, Ogretmen B.v Sphingosine kinase-1 and sphingosine 1-phosphate receptor 2 mediate Bcr-Abl1 stability and drug resistance by modulation of protein phosphatase 2Av. *Blood.* 2011 Jun 2; 117(22):5941-52.

[259] Baran Y, Salas A, Senkal CE, Gunduz U, Bielawski J, Obeid LM, Ogretmen B. Alterations of ceramide/sphingosine 1-phosphate rheostat involved in the regulation of resistanceto imatinib-induced apoptosis in K562 human chronic myeloid leukemia cells. *J Biol Chem.* 2007 Apr 13; 282(15):10922-34. Epub 2007 Feb 15.

[260] Marfe G, Di Stefano C, Gambacurta A, Ottone T, Martini V, Abruzzese E, Mologni L, Sinibaldi-Salimei P, de Fabritis P, Gambacorti-Passerini C, Amadori S, Birge RB. Sphingosine kinase 1 overexpression is regulated by signaling through PI3K, AKT2, and mTOR in imatinib-resistant chronic myeloid leukemia cells. *Exp Hematol.* 2011 Jun; 39(6):653-665.e6.

BIOGRAPHICAL SKETCHES

Gabriella Marfe is a researcher at Scienze e Tecnologie Ambientali, Biologiche e Farmaceutiche, Second University of Naples, Naples, Italy. Her basic research focuseson cancer resistance and with a special emphasis on colon and leukemia cancer stem cells.

Carla Di Stefano is a researcher at of Hematology, "Tor Vergata" University, Rome Italy. Her basic research focuses on leukemia cancer stem cells.

In: A Closer Look at Kinase Inhibition
Editor: Alex Tompson
ISBN: 978-1-53616-973-7
© 2020 Nova Science Publishers, Inc.

Chapter 2

ROLE OF KINASES IN ANTIVIRAL THERAPEUTICS

Vaishali M. Patil[1,*] ***and Neeraj Masand***[2]
[1]Drug Discovery and Interaction Center, KIET School of Pharmacy, KIET Group of Institutions, Ghaziabad, Uttar Pradesh, India
[2]Department of Pharmacy, Lala Lajpat Rai Memorial Medical College, Meerut, Uttar Pradesh, India

ABSTRACT

Kinases are a part of many cell processes and have been linked to some cancer treatment targets. Protein kinases represent central and multifunctional regulators of a balanced virus-host interaction. Host cell kinases are important for the replication of a number of hemorrhagic fever viruses. Cyclin-dependent kinases (CDKs) are required for replication of viruses that replicate only in dividing cells, non-dividing cells and plays crucial regulatory role in cell cycle, transcription, and replication of many viruses. Several kinase inhibitors developed as anti-cancer drugs, also possess ability to disrupt productive replication and dissemination in viral infected cells and thus exhibit anti-viral activity. Some of the kinase

* Corresponding Author's Email: vaishuwise@gmail.com.

inhibitors (ex. LDC4297, OSU-03012, WV970and BIBX 1382) have demonstrated inhibition of different viruses.

In addition to this, some studies highlight the multiple unexpected ways that viruses use host cell kinases during infection. Consequently, kinase inhibitors are attractive candidates for "repurposing" as broad-spectrum anti-viral agents.This chapter summarizes host cell and viral protein kinases, progress in the use and repurposing of approved kinase inhibitors as broad-spectrum antivirals.

Keywords: kinase inhibitors, host cell kinases, viral protein kinases, antiviral drugs, drug repurposing

INTRODUCTION

The governments and World Health Organization (WHO) are dealing with critical emerging/reemerging viruses having pandemic potential and have caused alarming outbreaks in recent years and lack specific treatments (example- Zika virus (ZIKV), Ebola virus (EBOV), Middle East respiratory syndrome coronavirus (MERS-CoV). In addition to this hundreds of human disease-causing viruses demand antiviral strategies. At present the approved antiviral drugs can treat around ten types of viral infections and similarly preventive vaccines are available. The currently available antiviral drugs belong to the category of direct-acting drugs which target individual virus proteins. Limitation with available direct-acting drugs is lack of spectrum of coverage to fulfill the clinical requirements. In USA, Food and Drug Administration (FDA) has approved only 12 new antiviral drugs from 2012 to 2017. Among them 8 are anti-hepatitis C virus (HCV) and 2 are anti-human immunodeficiency virus (HIV) drugs (www.fda.gov). With the emerging challenges in antiviral drug discovery such as the direct-acting approach, high cost and longer time required limit the drug development process. In addition to this, lack of global health protection policies against existing, resistance and/or newly emerging viruses is a challenge for the conventional approach. To overcome these limitations and to provide a possible solution the host-targeted broad-spectrum antiviral strategy can be an alternative approach.

Table 1. List of FDA approved anti-cancer drugs as kinase inhibitors

Drug target (Kinase)	Kinase inhibitor drugs
ALK	Crizotinib, Ceritinib, Alectinib, Brigatinib
BCR-Abl	Bosutinib, Dasatinib, Imatinib, Nilotinib, Ponatinib
B-Raf	Vemurafenib, Dabrafenib
BTK	Ibrutinib
CDK	Palbociclib, Sorafenib, Ribociclib
c-Met	Crizotinib, Cabozantinib
EGFR	Gefitinib, Erlotinib, Lapatinib, Vandetanib, Afatinib, Osimertinib
JAK	Ruxolitinib, Tofacitinib
MEK1/2	Trametinib
PDGFR α/β	Axitinib, Gefitinib, Imatinib, Lenvatinib, Nintedanib, Pazopanib, Regorafenib, Sorafenib, Sunitinib
RET	Vandetanib
Src	Bosutinib, Dasatinib, Ponatinib, Vandetanib
VEGFR	Axitinib, Lenvatinib, Nintedanib, Regorafenib, Pazopanib, Sorafenib, Sunitinib

In the past few decades, identification of kinases (enzymes involved in phosphorylation) has been a topic of interest. Approximately 538 kinases are encoded in the human genome. As phosphorylation is one of the major mechanisms to regulate the activities of protein factors, in mammalian cells it impacts various key-functions such as signal transduction, nucleo-cytoplasmic shuttling, macromolecular complexes assembly, DNA binding and regulation of enzymatic activities (Coussens et al., 1986; Manning et al., 2002; Fabbro et al., 2015). The various role of kinases in oncological issues includes- (i) carcinogenesis and metastases has been elucidated with the understanding of the molecular mechanisms involved in cancer cell signalling (Kostler et al., 2015);(ii) protein kinases promote cell proliferation, survival and migration (Maurer et al., 2011); (iii) cancer initiation, promotion, progression and recurrence (Kostler et al., 2015; Kittler et al., 2018). In addition to oncological issues, dysregulation of kinases has been demonstrated in many human disorders including immune, neurological and infectious diseases (Sato et al., 2005; Mueller et al., 2005; Chong et al., 2012; Tabit et al., 2013). Encouraging results are observed during the current development of robust and selective kinase inhibitors with

>10,000 patent applications in USA alone, effective kinase-targeted antibodies, and 35 FDA approved drugs. Some of the FDA approved anti-cancer kinase inhibitors are listed in Table 1 and chemical structures in Figure 1.

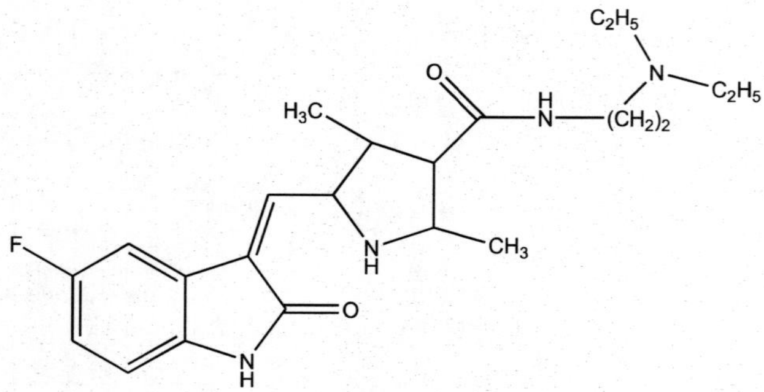

Sunitinib

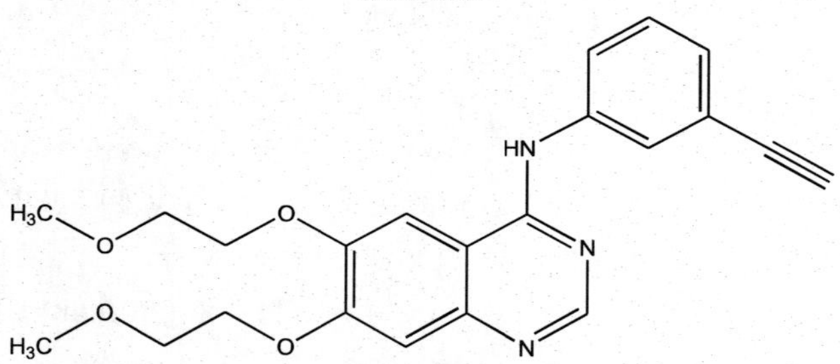

Erlotinib

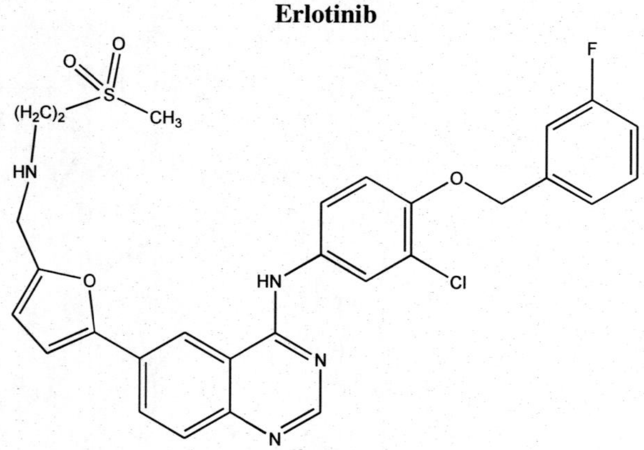

Lapatinib

Figure 1. (Continued).

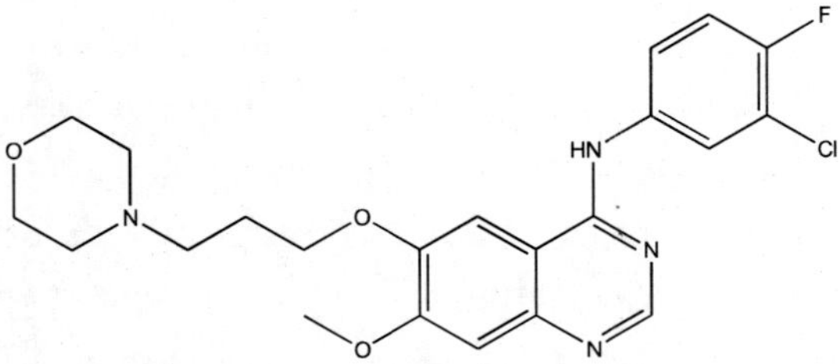

Gefitinib

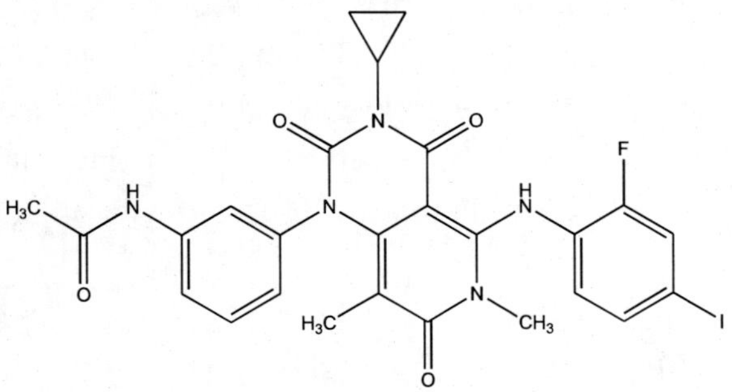

Trametinib

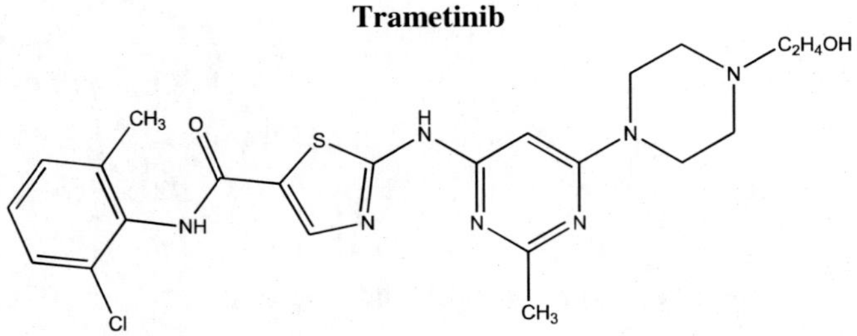

Dasatinib

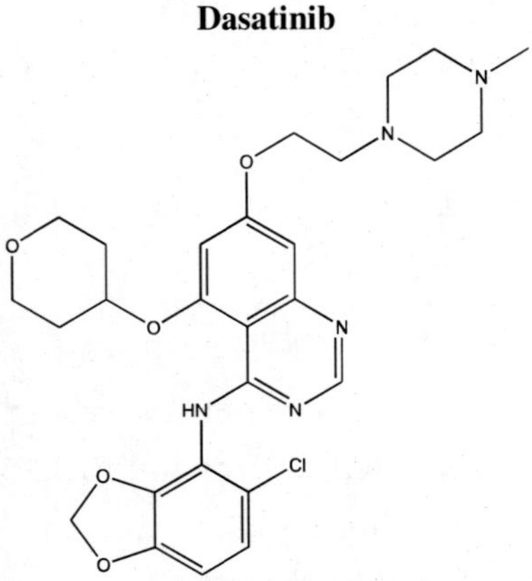

Saracatinib

Figure 1. Chemical structures of kinase antiviral agents.

The infection causing viruses may rely on the viral protein phosphorylation for replication, propagation and other diverse steps of their life cycle. The protein kinase activity by particles from diverse virus families has been reported. Large DNA viruses encode for viral kinases, RNA viruses (retroviruses) can hijack the signaling machinery of the host cell and embark cellular kinases when budding (ex. HIV-1) (Giroud et al., 2011). This chapter covers details on host-cell protein and viral protein kinases, their function in viral life cycle, various antiviral kinase inhibitors and the strategies for kinase drug repurposing strategies in antiviral drug development. The strategies developed towards identification of viral therapeutic targets and their inhibitors are discussed with suitable examples.

Host-Cell Protein Kinases

Host-cell kinase inhibitors offer great potential to apply antiviral drug repurposing strategies as viruses target a large number of host kinases at various steps during their life cycle (Supekova et al., 2008; Li et al., 2009; Keating and Striker, 2012; Jiang et al., 2014). Some of the host kinases are essential and thus offer attractive target for development of broad-spectrum antivirals (Schor and Einav, 2018). These host cell kinases along with reported inhibitors are further discussed in detail.

Numb-Associated Kinases (NAK)

The family of Ser/Thr kinases, Numb-associated kinases (NAK) is a promising target for development of broad-spectrum antiviral kinase inhibitors. The host kinases adaptor protein 2 (AP2)-associated protein kinase 1 (AAK1) and cyclin G-associated kinase (GAK) is involved in regulation of viral entry, assembly, release of unrelated viruses. Sunitinib and erlotinib, the approved anticancer drugs can inhibit AAK1 and GAK, respectively along with inhibition of dengue virus (DENV) and EBOV (Neveu et al., 2012, 2015; Bekerman et al., 2017). The possible mechanism of action is inhibition of AP-mediated intracellular membrane trafficking regulated by AAK1 and GAK. Both these kinase inhibitors have shown protection against morbidity and mortality when used in combination in murine models of DENV and EBOV infection (Bekerman et al., 2017).

Epidermal Growth Factor Receptor (EGFR)

EGFR, a type of receptor tyrosine kinase is an essential host factor required by viruses at various stages during their life cycle (Zheng et al., 2014) and has been validated as a broad spectrum antiviral target. In human cytomegalovirus (HCMV) and adeno-associated virus subtype 6 (AAV6), it is required as an entry coreceptor (Wang et al., 2003; Weller et al., 2010). In HCV and influenza A virus (IAV), it mediates signaling during endocytosis (Eierhoff et al., 2010; Lupberger et al., 2011; Diao et al., 2012; Neveu et al., 2012, 2015; Zona et al., 2013). During postentry, EGFR regulates F-actin dynamics through nuclear trafficking (ex. HCMV) (Wang et al., 2005). Viruses like IAV target epidermal growth factor receptor (EGFR) to evade host immune system (Ueki et al., 2013). The approved anticancer kinase inhibitors target EGFR (ex. erlotinib and gefitinib) and ERBB2 (ex. Lapatinib) during *in vitro* antiviral studies against HCV (Lupberger et al., 2011; Diao et al., 2012; Neveu et al., 2015), HCMV (Schleiss et al., 2008), and poxvirus (Langhammer et al., 2011).

MAPK/ERK

The mitogen-activated protein kinases/extracellular signal-regulated kinases (MAPK/ERK) are essential for viral replication and thus their inhibitors have potential to target viruses. ERK1/2 phosphorylates HIV-1 proteins for enhanced viral infectivity (Yang and Gabuzda, 1999). IAV requires this pathway for virus production, export viral ribonucleoprotein complexes from the nucleus through its activation (Pleschka et al., 2001). In EBOV, MAPK/ERK pathway is used for disease pathogenesis by mediating entry into host cell, improved cytokine production, and cellular toxicity (Zampieri et al., 2007; Kolokoltsov et al., 2009; Johnson et al., 2014). The anticancer kinase inhibitors, trametinib (MEK1/2) and selumetinib (MEK1/ERK1/2) have shown inhibition of this pathway in Middle East respiratory syndrome coronavirus (MERS-CoV) (Kindrachuk et al., 2015).

Src Family of Kinases (SFKs)

In the life cycle of various viruses Src kinases play an important role (Pagano et al., 2013). SFKs like Fyn, Lyn, c-Src, Csk (DENV), c-Yes (West Nile Virus, WNV), and Csk (HCV) regulate replication of viruses from family Falviviridae. (Hirsch et al., 2005; Chu and Yang, 2007; Supekova et al., 2008; de Wispelaere et al., 2013; Kumar et al., 2016). The HIV-1 also activates SFKs to effectively replicate (Narute and Smithgall, 2012). The SFK inhibitors, dasatinib (FDA approved) and saracatinib (Phase III) inhibit DENV assembly *in vitro*. The antiviral activity i.e., inhibition of DENV RNA replication was found to be Fyn-mediated (de Wispelaere et al., 2013).

Abelson tyrosine kinase (Abl)

Some reported kinase inhibitors, dasatinib and saracatinib target cellular Abelson tyrosine kinase (c-Abl). C-Abl is a member of the host kinase from Abl family of nonreceptor tyrosine kinases. C-Abl is implicated in the budding or release (poxviruses, EBOV) (Reeves et al., 2011; Garcia et al., 2012), virus entry (coxsackievirus, polyomavirus, DENV) (Coyne and Bergelson, 2006; Swimm et al., 2010; Clark et al., 2016). The approved anticancer compounds, imatinib and nilotinib are c-Abl inhibitors and fail to inhibit SFK. Both of them show inhibition of replication in EBOV, DENV, MERS-CoV, SARS-CoV (Garcia et al., 2012; Dyall et al., 2014; Clark et al., 2016). Dasatinib also inhibited MERS-CoV and SARS-CoV (Dyall et al., 2014). Imatinib also reduces viral load, spread and mortality in a murine model of vaccinia virus (Reeves et al., 2011) but in *in vivo* mice model it induces immunosuppression without protection. For Src and Abl inhibitors, the proposed mechanism of viral inhibition is through inhibition of actin motility (Reeves et al., 2011). The molecular targets for these inhibitors contributing for antiviral activity need to be investigated in animal models of other viral infections.

Cyclin-Dependent Kinases (CDKs)

The host cyclin-dependent kinases (CDKs) family is an important target for various viruses to regulate the host cell cycle through modulation of expression and activity (Schang, 2003). As an important antiviral target, inhibition of CDKs lead to suppression of viral replication along with suppression of viral persistence and/or transformation is reported (Li et al., 2003). In adenovirus and hepatitis B virus (HBV), the replication is promoted through progression into late-G1/S-phase by increased CDK2 phosphorylation and increased activity of the G1-phase kinase CDK-4, respectively (Gearhart and Bouchard, 2010; Cheng et al., 2013). While in HIV-1, both CDK-6 and -9 have been validated as potential antiviral targets (Nemeth et al., 2011; Pauls et al., 2014b). In herpes simplex virus type 1 (HSV-1) infection, CDK-9 is the potential target (Durand and Roizman, 2008). Various CDK inhibitors with variable selectivity have demonstrated antiviral activity (Figure 2) (Chao et al., 2000; Schang et al., 2006; Guendel et al., 2010; Yamamoto et al., 2014; Pauls et al., 2014a; Perwitasari et al., 2015; Xu et al., 2016; Badia et al., 2016).

Studies have reported the potential mechanism of action (MOA) underlying the antiviral effect of CDK inhibitors as-

(i) inhibition of phosphorylation of viral proteins (IAV and varicella zoster virus), (Habran et al., 2005; Perwitasari et al., 2015);
(ii) reduced intracellular dNTP availability and viral mRNA transcription through inhibition of a cellular protein phosphorylation (ex. Palbociclib exerts its anti-HSV-1 and HIV-1 activities) (Pauls et al., 2014a; Badia et al., 2016).

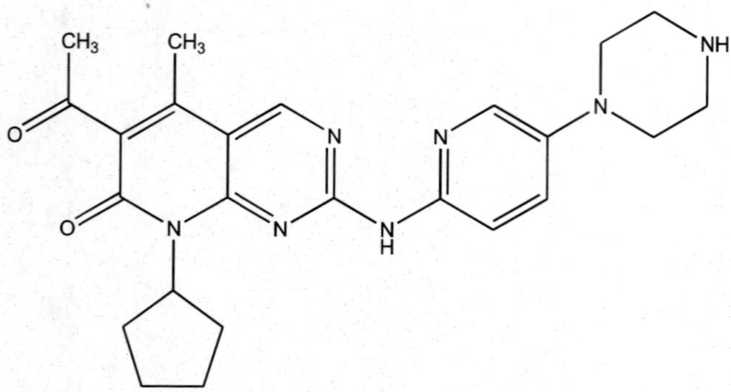

Palbociclib (Approved CDK4/6 inhibitor) Suppressed HIV-1 reverse transcription and HSV-1 replication *in vitro*

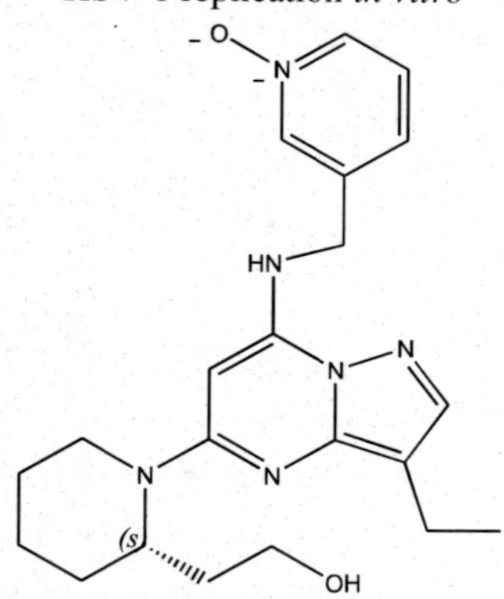

Dinaciclib (SCH 727965)(CDK-1/2/5/9 inhibitor, Phase III)

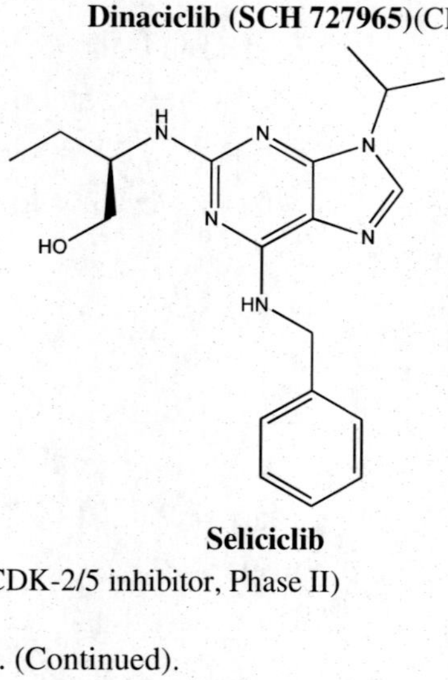

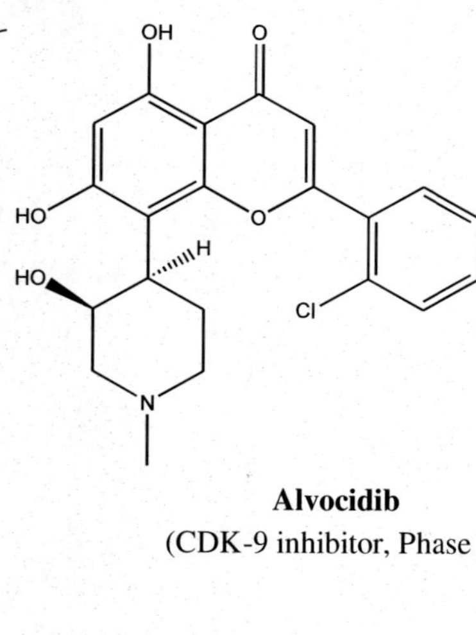

Seliciclib
(CDK-2/5 inhibitor, Phase II)

Alvocidib
(CDK-9 inhibitor, Phase II)

Figure 2. (Continued).

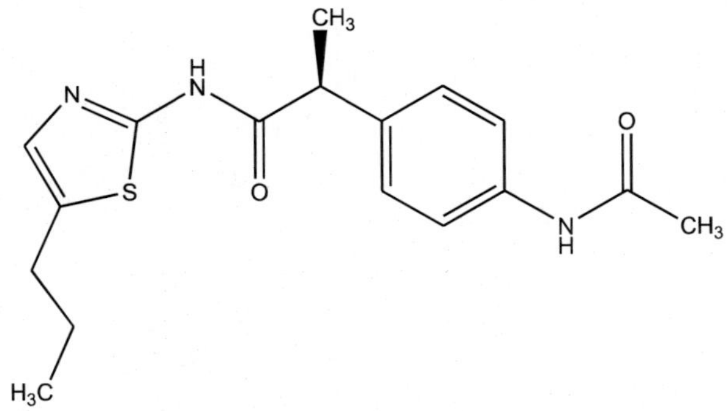

PHA-690509 (CDK-2 inhibitor, Phase I)

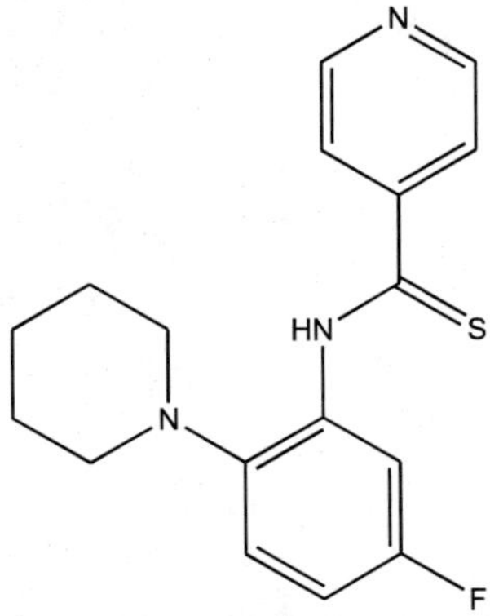

FIT-039(HSV-1 inhibitor, *in vitro* activityagainst DNA viruses)

Figure 2.Details of some CDK inhibitors.

Phosphatidylinositol 3¢kinase–Akt (mammalian)

Targeting the PI3K/Akt/mTOR pathway multiple viruses activate the phosphatidylinositol 3¢kinase–Akt–mammalian target of rapamycin (PI3K/Akt/ mTOR) pathway to support their replication and escape cellular stress responses (Buchkovich et al., 2008). Similarly, HCV, WNV and IAV promote various steps in life cycle and to control apoptosis through activation of this pathway (Mannova´ and Beretta, 2005; Ehrhardt et al., 2006; Hale et al., 2006; Shives et al., 2014). Some of the compounds evaluated as Akt inhibitors are listed in Figure 3.

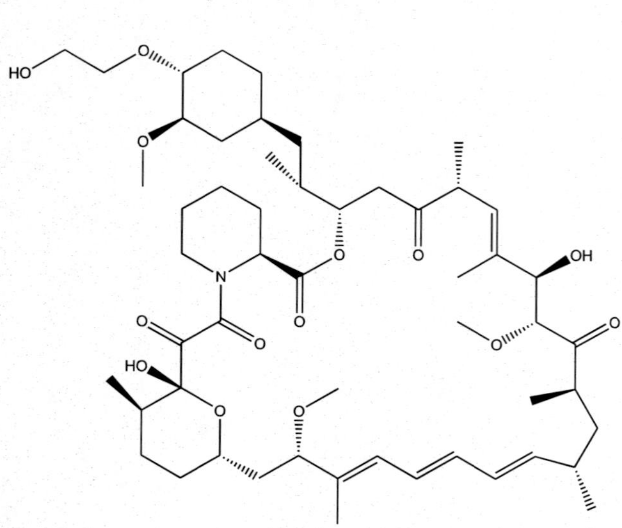

Everolimus

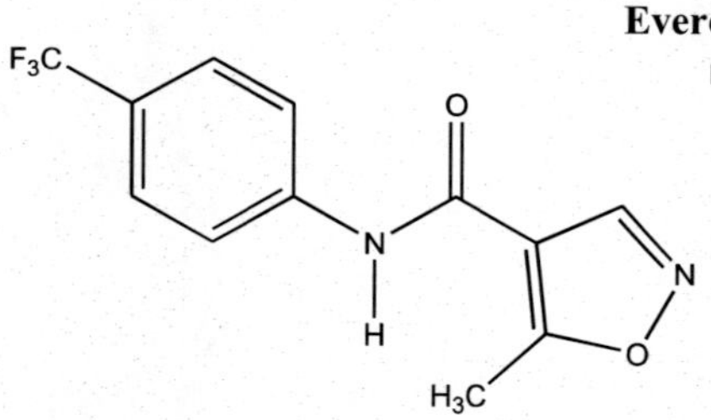

Leflunomide

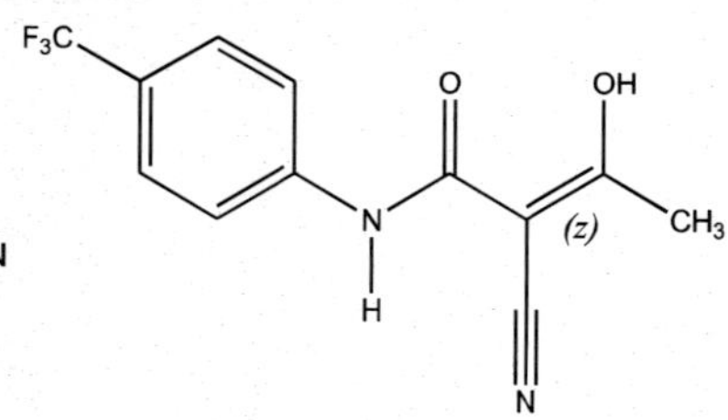

Teriflunomide

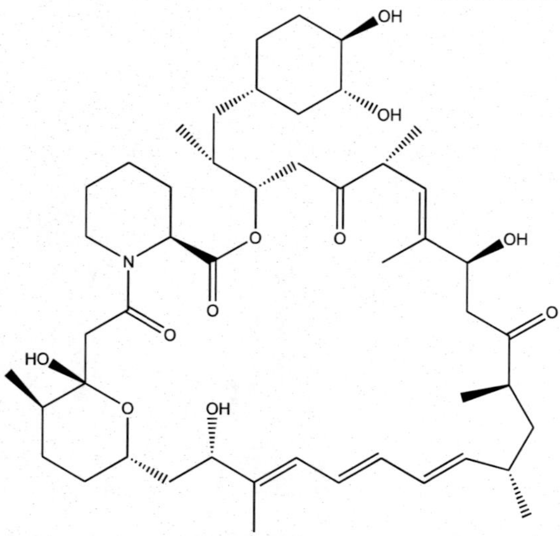

Sirolimus

Figure 3.Chemical structures of Akt inhibitors.

In cells infected with MERS-CoV, this pathway was shown to be activated by everolimus (mTor inhibitor). It is an approved anti-cancer agent to prevent transplant rejection. It has only mild *in vitro* activity against MERS-CoV. The approved AKT inhibitor, miltefosine fails to show activity in this infection model (Kindrachuk et al., 2015). Teriflunomide, an active metabolite of leflunomide has suppressed Akt phosphorylation. It is approved for rheumatoid arthritis treatment and has shown potent anti-HCMV activity (*in vitro* and *in vivo* model) (Waldman et al., 1999). In addition to this, single drug or combination treatment of leflunomide and everolimus/sirolimus (approved immunosuppressant mTOR inhibitors) reduces BK viremia and disease progression in patients with BK virus nephropathy (Williams et al., 2005; Jung et al., 2013; Jaw et al., 2017). The proposed mechanism of action involves direct suppression of polyomavirus replication (Liacini et al., 2010). These observations support the importance of further exploring the broad antiviral activity of these classes of kinase inhibitors.

Viral Protein Kinases

The major mechanism involved to regulate the activity of protein factors is phosphorylation. Up to 30% of all proteins in mammalian cells are modified by phosphorylation and such regulation impacts multiple levels, including nucleo-cytoplasmic shuttling, the assembly of macromolecular complexes, DNA-binding capacity and enzymatic activation. The presence of kinase activities in viral particles has been reported and earlier observations have demonstrated high levels of protein kinase activity are packaged in certain viruses (Rauscher murine leukemia virus, vaccinia virus, Rous sarcoma virus) (Strand et al., 1971; Paoletti et al., 1972; Hatanaka et al., 1972). These findings have developed interest in the study of virus/kinase relationships. Protein kinases encoded by the viral genome catalyze the phosphorylation of viral proteins in large DNA viruses (Herpesviridae, Poxviridae, Baculoviridae) (Jacob et al., 2011).

Further the protein kinases can be characterized by their ability to get phosphorylated and interfere with the enzymatic activity, cellular location, and association with other proteins resulting in to changes in the functional group of target protein. The protein kinases can be categorized according to the specificity of the substrate amino acids (ex. serine (Ser), threonine (Thr), tyrosine (Tyr) kinases). The consensus sequence of protein kinases can phosphorylate Ser, Thr or Tyr residue. The catalytic core of protein kinases consists of a common structure of 12 conserved subdomains and these are essential for kinase activity such as ATP binding. In human, the 500 protein kinase genes are present; regulate the cellular pathways, viral replication and plethora of different functions (Hanks and Hunter, 1995; Manning et al., 2002). In 1970s, the existence of viral proteins has been reported followed by their identification (Tan, 1975; Collet and Erikson, 1978; Leader and Katan, 1988; Leader, 1993). The viruses encode several protein kinases is supported by the fact that, viruses have complex interaction with the host proteins and mimic cellular proteins to use the cellular machinery.

Viral Ser/Thr and Tyr protein kinases have been reported. Viral Ser/Thr kinases show less homology with other reported cellular kinases and are encoded by large and evolutionarily old DNA viruses (herpesviruses, poxviruses, baculoviruses, and other virus families like asfavirus, nimavirus, phycodvirus, iridovirus, mimivirus) (Tidona and Darai, 1997; Jakob et al., 2001). HBV and vesicular stomatitis virus (VSV) may encode structurally different protein kinases (Leader, 1993). Protein phosphatases are encoded by vaccinia virus and bacteriophage lambda and play essential role in viral infection like cellular protein phosphatases. The function of many of these viral enzymes represent possible new targets for anti-viral therapy (Leader, 1993).

The viral tyrosine kinases is the best characterized oncogenic kinase encoded by acute transforming retroviruses (Maeda et al., 2008). Among the reported three subfamilies (α, β and γ) and eight human herpesviruses (HSV-1, HSV-2, VZV, HCMV, HHV-6, HHV-7, EBV, KSHV), all encode Ser/Thr kinases. The orthologs of conserved herpesvirus protein kinases (CHPK) include UL13 of HSV, ORF47 of VZV, UL97 of HCMV, U69 of HHV-6, BGLF4 of EBV, and ORF36 of KSHV and murine herpesvirus 68 (MHV-

68). The Ser/Thr kinases have been described (ex. UL13 kinase and US3 kinase of HSV) in the large subunit of the ribonucleotide reductase of HSV-2. The viral kinases have been encoded for distinct members of the herpesviruses and they have aided in the adaptation of the virus to the host complexity (Aoki et al., 2007). In VZV, CHPK kinase does not affect viral growth in cell culture but is essential for replication in human T cells and human thymus/liver xenografts (Moffat et al., 1998). The kinases are reported in poxviruses (B1 kinase, F10 kinase) and in Baculoviruses (PK1).

Viral Ser/Thr protein kinases are promising antiviral drug targets as they play essential role in virus virulence and are distinct from the cellular protein kinases to facilitate drug selection without interfering with the cellular kinase activity. One of the successful example is maribavir with anti-HCMV activity as inhibitor of UL97 kinase which passed phase I and II clinical trials and failed during phase III studies (Figure 4) (Littler et al., 1992; Sullivan et al., 1992; Biron et al., 2002; Trofe et al., 2008). In addition to this, for poxvirus replication cycle, B1 kinase is essential and has been evaluated by co-administration with cidofovir, an antiviral drug (Figure 5) (Vigne et al., 2009).

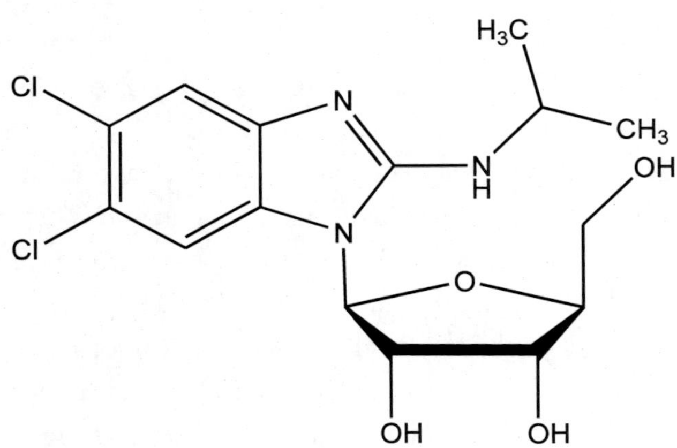

Figure 4.Chemical structure of Maribavir.

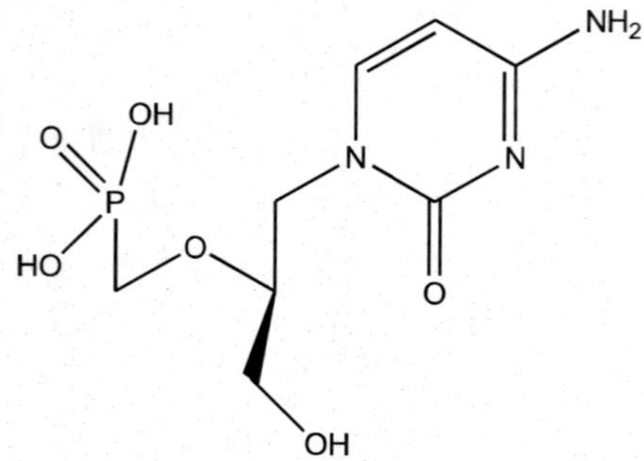

Figure 5.Chemical structure of Cidofovir.

pUL97

The human cytomegalovirus (HCMV) is characterized by the tight virus-host cell interaction during replication and the protein kinase pUL97 is an important determinant of virus replication. Its kinase activity contributes towards nuclear lamina disassembly during the late phase of replication (Graf et al., 2013). Due to structural and functional characteristics, the HCMV-encoded pUL97 is a cyclin-dependent kinase (CDK) ortholog. Some studies have reported pUL97 bridging to the core nuclear egress complex (NEC) (Marschall et al., 2005). The pUL97 interacts with a subset of cyclins (cyclin T1, cyclin B1, etc.). Cyclin B1 interacts at higher affinity and is dependent on pUL97 activity (Steingruber et al., 2015). The involved mechanism for pUL97-cyclin B1 interaction is determined by active pUL97 domain based on the loss of activity in the presence of pUL97 inhibitors or mutational changes in the conserved kinase domain or the non-conserved C-terminus of pUL97. Further, mass spectroscopic approach was used to analyze cyclin interaction of pUL97 and other viral proteins and has been analyzed using *in vitro* kinase assay. The studies show differential affinities for human cyclins. Maribavir (MBV) inhibits only pUL97 and cyclin B1 interaction. In HCMV-infected cells, cyclin H interacts at high-affinity. In cyclin-associated complexes, all known pUL97 substrates (i.e., viral phosphoproteins) can be detected (Steingruber et al., 2016).

In a recent study, the association of viral capsids with the HCMV-specific core NEC and pUL97 has been analyzed showing direct association of nuclear HCMV capsids with NEC proteins at viral budding sites of the nuclear lamina. Intranuclear interaction was observed between virions and the core NEC (pUL53) and pUL97. During *in vitro* assay, kinase activity was observed for capsid associated pUL97. The pUL97 is a component of the herpesvirus-characteristic tegument layer and is packaged into HCMV particles which are released from the host cell (Kalejta, 2008; Prichard, 2009). Thus direct targeting of HCMV-pUL97 by kinase inhibitors is a novel antiviral approach (Milbradt et al., 2018).

DENV and Protein Kinase C

The role of Casein Kinase I (CK I) or protein kinase G (PKG) to phosphorylate NS5 at the methyltransferase domain has been reported (Bhattacharya et al., 2009a and b). For the normal functioning of NS5 the phosphorylation is essential as blocking the phosphorylation by chemical inhibitor or amino acid substitution of the potential phosphorylation site suppresses viral replication and production (Bhattacharya et al., 2009b). In addition to this, the reports suggested that NS5 is a central protein responsible for mediating functional interactions between host and virus proteins. Detailed study of NS5 phosphorylation is required for better understanding of the viral life cycle as well as to provide novel approaches based on the viral-host interaction. Noppakunmongkolchai et al., (2016) performed in silico analyses to identify human kinases regulating DENV2 NS5 by studying the various amino acid sequences. The functional interaction between NS5 and PKC were predicted by the phosphor-algorithms (Scansie, NetPhosk 2.0 and KinasePhos 2.0) and identified potential phosphorylaion sites (Thr244, Thr320, Ser796 and Ser885). The vector-borne flaviviruses conserves Thr302 and Ser796 while the rest two are conserved within DENVs. In NS5, the methyltransferase domain contains Thr244 and RdRp domain contains Thr302, Ser796 and Ser885. These possible PKC phosphorylation sites on the NS5 were confirmed by *in vitro* kinase assay. Further studies are required to evaluate the possibility of Thr244 as phosphorylation site by PKC. Further *in vivo* studies on the functional interaction between host PKC and DENV NS5 have shown that silencing of the PKC decreases Ser and Thr phosphorylations of NS5. During inhibition of host cell PKC by bisindolylmaleimide I (BisI), the blockage was observed (Figure 6). It concluded that PKC activity and DENV entry into host cells are not correlated to each other while after 12 hr of BisI treatment, PKC inhibition promotes viral replication and thus PKC stimulation suppresses DENV replication in host cells.

HIV-1 Associated Kinases

In the ritroviridae family, ERK2, PKA and NDR1/2 have received major interest to define their incorporation and functions. HIV-1-associated kinases and their possible function in the retroviral life cycle have been demonstrated and complete understanding is required to develop new and relevant drug discovery strategies (Giroud et al., 2011). Some of the virion-associated kinases reported in HIV-1 and simian immunodeficiency virus (SIV) are ERK2 and PKA and are packaged into particles (Hansen et al., 1991; Cartier et al., 1997 and 2003; Jacque et al., 1998; Stephenson et al., 2011). In HIV-1, ERK2 (MAPK) is the first identified protein kinase in addition to protein kinase C (PKC), phosphoglycerate kinase 1 and p56 tyrosine kinase (Cartier et al., 1997; Jacque et al., 1998; Ott et al., 2000; Chertova et al., 2006). Studies have been reported based on kinase-deficient viruses, mutant viruses or use of chemical inhibitors of cellular kinases (inhibitor-based approach). ERK2 enhances replication in HIV-1 protein (Vif) and poor infectious properties have been observed with HIV-1 particles devoid of ERK2 activity (Jacque et al., 1998; Yang et al., 1999). According to the inhibitor-based approach, drug/inhibitor may abolish the virus-associated enzymatic activity and interfere with the cellular kinases required for viral replication. The mechanism involved in the incorporation of cellular kinases into the viral particles can be investigated by inhibition of protein-protein or protein-nucleic acid interactions. The key regulators of HIV-1 replication are phosphorylation and dephosphorylation and the virus can hijack cell signaling system to fulfill respective functions at the site of replication. Some of the anti-HIV-1 strategies are- targeting HIV-1 associated cellular proteins, and inhibition of intracellular steps of HIV-1 replication (Devaux et al., 1990 and 1999; Hansen et al., 1991; Arthur et al., 1992; Briant et al., 1999; Tardif et al., 2009). Some of the examples include purine derivative, Roscovitine which inhibits cyclin-dependent kinase (CDK) required for RNA-polymerase II dependent transcription (Schang et al., 2006). Thus the careful examination of virus associated kinases is necessary to determine their role as target in viral infected cells.

KINASES AND THEIR POSSIBLE FUNCTIONS

Understanding the molecular events accompanying virus replication has been a major focus of experimental virology, and is essential for the proper understanding and control of all virus diseases. The biological "purpose" of any replication cycle is the generation of new viral genomes and proteins in sufficient quantities to ensure propagation of the viral genome; this requires protection of the extracellular viral genome from enzymatic degradation and can be introduced into target cells for further rounds of replication. Much has become known about the initial stages of attachment, and more detailed study shows that the initial recognition between virus and host is more complex. Temporal regulation of intracellular events is critical in all but the very simplest of viruses, with some form of suppression of the host innate immune response being common to nearly all human viruses. There are examples where the innate immune response is even used to enhance virus spread to cells otherwise unavailable to the virus. Over the next few years, the boundaries between virus-directed events and cellular processes that control specialized cell functions are likely to be even more complex; nevertheless understanding these processes will open up a range of targets for the development of novel antiviral therapies and immunotherapy. Some of the cellular protein kinases present in viruses from retroviridae family and their functions are summarized in Table 3.

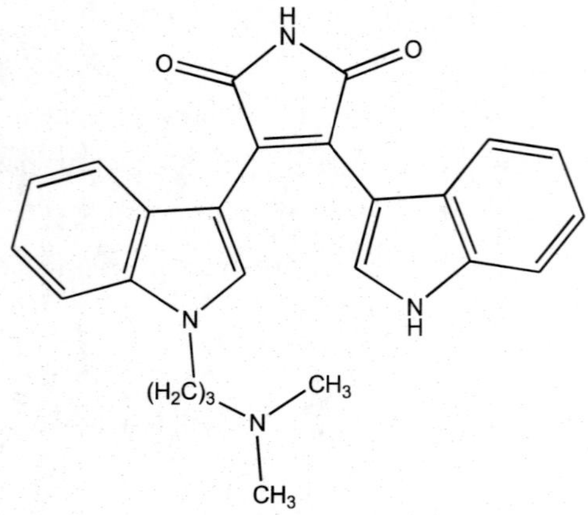

Figure 6.Chemical structure of Bisindolylmaleimide I (BisI).

Table 3. Cellular protein kinases and their functions from family Retroviridae

Virus	Virus-associated cellular kinases	Function/s	Ref.
AMV	42-46 kDa and 60-64 kDa kinases	10 to 25 kDa viral potein	Strand and August, 1971; Tsiapalis, 1977; Houts et al., 1978
RSV	Cellular kinase	-	Hatanaka et al., 1972
MMTV	Cellular kinase	-	Hatanaka et al., 1972
MSV	Cellular kinase	-	Sen and Todaro, 1979; Sen et al., 1979; Sen, 1981
R-MLV	42-46 kDa and 60-64 kDa kinases	-	Strand and August, 1971; Tsiapalis, 1977; Houts et al., 1978
FelV	Cellular kinase	-	Hatanaka et al., 1972
HTLV-1	ERK2	Virus assembly and release	Cartier et al., 1997; Hemonnot et al., 2006
HIV-1	ERK2, PKA, DR1/2, 53 kDa, p56, PKC, PRP2, Nm23-Ha	Virus infectivity, uncoating, Virus release, replication	Chertova et al., 2006; Jacque et al., 1998; Hemonnotet al., 2004
SIV	ERK2, PKA, PKC	-	Bukong et al., 2010; Stephenson et al., 2011
FIV	ERK2	-	Bukong et al., 2010

KINASE INHIBITORS AS ANTIVIRAL AGENTS

Various kinases having proven role in the viral life cycle have been targeted in an effort to develop antiviral drugs. Some of the major highlights are discussed here. The protein kinase pUL97 is an important determinant of virus replication. Recently, indolocarbazoles were specified as a class of substances that inhibit the pUL97 kinase activity *in vitro*. In parallel, it was shown that indolocarbazoles interfere with HCMV replication; however, the causal relationship between inhibition of pUL97 kinase activity and virus replication need to be demonstrated. Here evidence is provided that indolocarbazole-mediated inhibition of virus replication is a direct result of diminished pUL97 protein kinase activity. In cell culture infections, a strong and selective antiviral activity was measured with respect to several strains of HCMV in contrast with other related or non-related viruses. For quantification, recombinant HCMVs expressing green fluorescent protein were used, demonstrating the high sensitivity towards compounds NGIC-I and Go$ 6976.

Interestingly, aganciclovir-resistant virus mutant (UL97-M460I) showed increased sensitivity to both compounds. Supporting this concept, transfection experiments with cloned pUL97 revealed that ganciclovir resistant mutants were characterized by reduced levels of auto-phosphorylation compared with wild-type and possessed especially high sensitivity to indolocarbazoles. Moreover, the Epstein– Barr virus-encoded homologous kinase, BGLF4, which showed a similar pattern of auto-phosphorylation and ganciclovir phosphorylation activities, was not inhibited. Importantly, acytomegalovirus deletion mutant, lacking a functional UL97 gene and showing a severe impairment of replication, was completely insensitive to indolocarbazoles. Thus, our findings indicate that a specific block in the activity of pUL97 is the critical step in indolocarbazole-mediated inhibition of virus replication and that pUL97 might be targeted efficiently by a novel antiviral therapy.

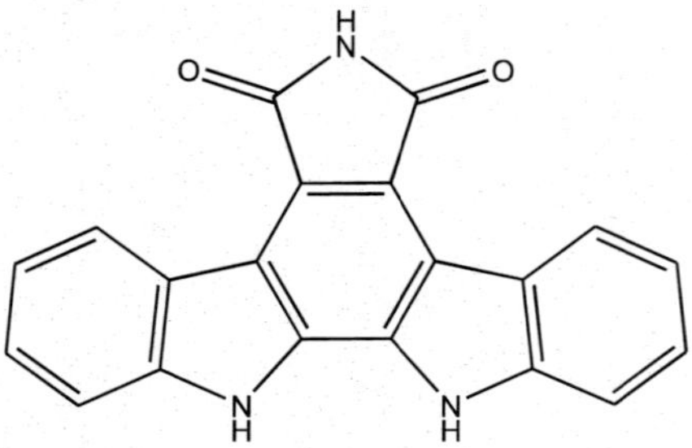

Figure 7.Chemical structure of Arcyriaflavin A.

The mechanism of kinase inhibition of antiviral drugs has been demonstrated with the help of some examples like selective anti-HCMV compounds from the chemical class of indolocarbazoles (Slater et al., 1999). The natural product Arcyriaflavin A is a potent inhibitor of HCMV replication in cell culture (Figure 7). A series of analogues (symmetrical indolocarbazoles) were synthesized to investigate structure activity relationships in this series against a range of herpes viruses (HCMV, VZV, HSV1, and 2). This identified a number of novel, selective and potent inhibitors of HCMV. Compound 12,13-dihydro-2,10-difluoro-5H-indolo[2,3-a]pyrrolo[3,4-c]carbazol e-5,7-(6H)-dione being the best example (IC_{50} = 40 nM, therapeutic index > 1450). Other compounds from this series were poor inhibitors of protein kinase C betaII, and failed to establish any correlation between their ability to inhibit HCMV and the enzyme PKC (Slater et al., 1999).

Later in 2000, the effect of indolocarbazoles was tested on a panel of viruses in the conventional plaquer eduction assay in the host cell type with high flexible for lytic virus replication. In genetically different HCMVs, treatment with NGIC-I or Go$ 6976 (50 nM) led to significant reduction in the replication (Zimmermann et al., 2000). Both the compounds failed to inhibit HSV-1 and human influenza A virus (strainA}WSN}33) and found to possess selectivity/specificity for HCMV by targeting the viral protein kinase. Further studies are required on pUL97-specific inhibitors as possible therapeutics for HCMV treatment as it is an important determinant of viral replication. As demonstrated by Herget et al., pUL97-specific kinase inhibitors can control HCMV replication using quinazoline compounds

(IC_{50} = 30 to 170 nM) (Herget et al., 2004). Quinazolines have shown inhibition of clinical isolates of HCMV resistant to ganciclovir and cidofovir-resistant viruses without development of resistant. The mechanism of action involves inhibition of pUL97 protein kinase activity. The role of pUL97 has been recently confirmed by LDC4297 acting as HCMV inhibitor.

LDC4297((*R*)-*N*-(2-(1*H*-pyrazol-1-yl)benzyl)-8-isopropyl-2-(piperidin-3-yloxy)pyrazolo[1,5-a][1,3,5]triazin-4-amine) is the first CDK7 inhibitor from the chemical class pyrazolotriazines with broad spectrum antiviral activity at nanomolar concentration (IC_{50} = 0.13 ± 0.06 nM) (Figure 8) (Hutterer et al., 2015). It has been analyzed for antiviral activity against viruses from different families like Herpesviride (HCMV, GPCMV, MCMV, HHV-6A, HSV-1/2, VZV, EBV), adenoviridae (HAdV-2 (type 2)), Poxviridae (Vaccinia virus), Retroviridae (HIV-1), and Orthomyxoviridae (Influenza A virus). In particular to HCMV, pUL97 regulates phosphorylation and inactivates Rb (Chou et al., 2006; Hume et al., 2008; Prichard et al., 2008; Gill et al., 2009; Kamil et al., 2009; Marschall et al., 2011). In viral CDK orthologs, the herpesviral UL-type (HvUL) protein kinases can phosphorylate Rb to conserve their function (Hume et al., 2008; Prichard et al., 2008; Rechter et al., 2009; Kuny et al., 2010). LDC4297 has shown suppression of high levels of phosphor-Rb and thus shows direct interference with the pUL97-mediated Rb phosphorylation.

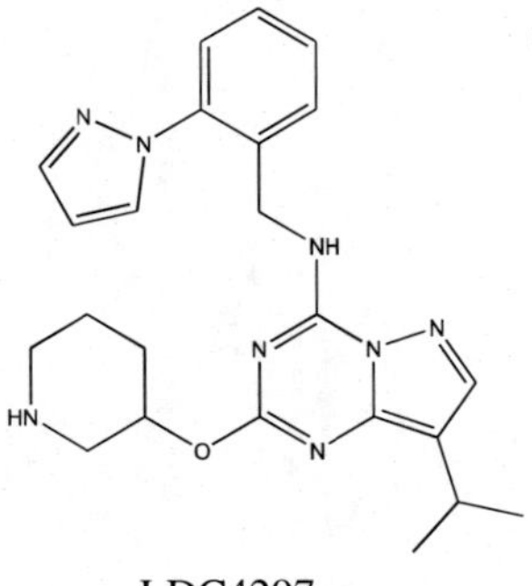

LDC4297

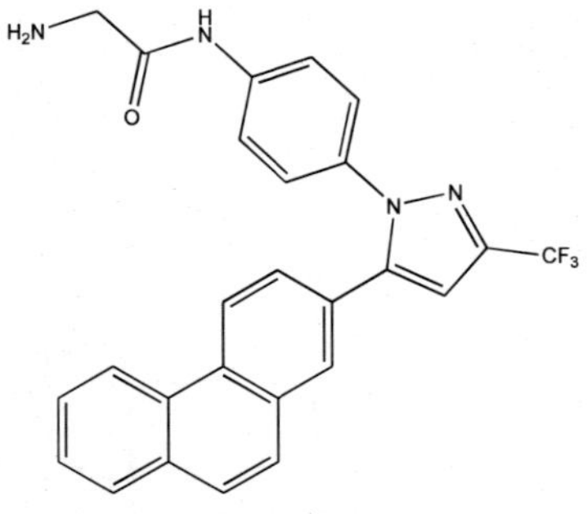

OSU-03012 (AR-12)

Figure 8. (Continued).

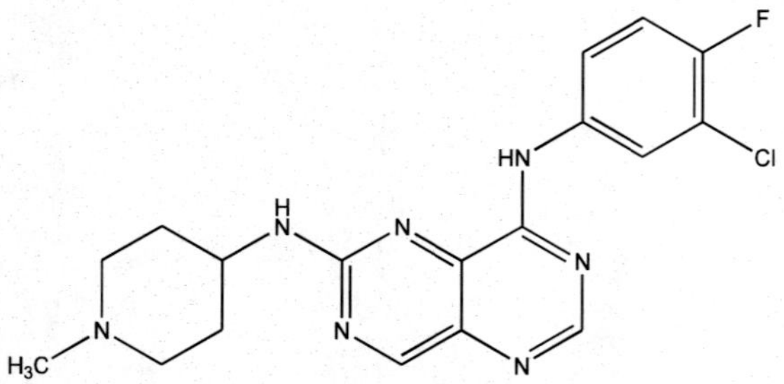

BIBX 1382

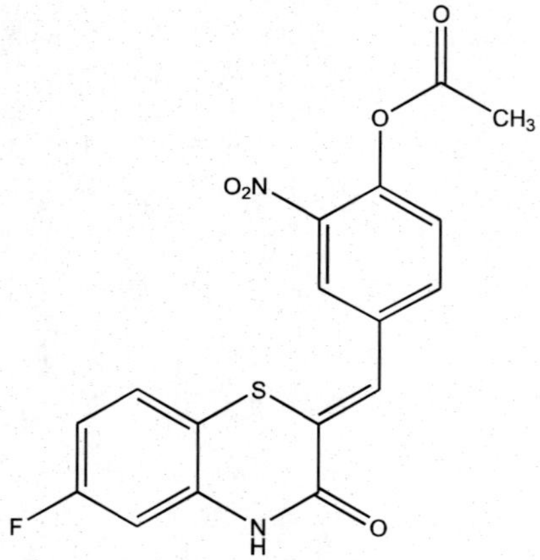

ON108110

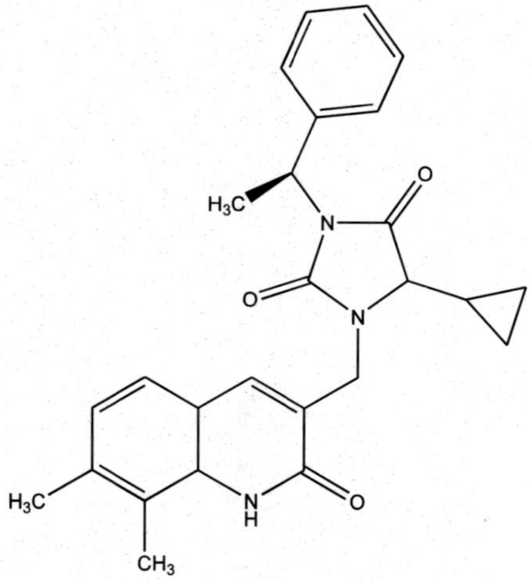

WV970

Figure 8. Chemical structures of antiviral kinase inhibitors.

PDK-1 is a Ser/Thr kinase and is activated by the phosphoinositide-3 kinase (PI3-K) mechanism in response to insulin and growth factor treatment (Mora et al, 2004). It possess the ability to phosphorylate and activate AKT, p70 S6 kinase, and possibly protein kinase C (PKC) isozymes (Alessi et al, 1997 and 1998; Stokoe et al, 1997; Le Good et al, 1998; Balendran et al, 2000). OSU-03012, a novel small molecule compound targeting PDK-1 pathway has been reported as anticancer agent (Figure 8) (Zhu et al, 2004; Cen et al., 2007). Later this celecoxib derivative, OSU-03012 (AR-12) is demonstrated to possess potent inhibitory properties against recombinant phosphoinositide-dependent kinase 1 (PDK-1) (IC_{50} = 5 μM). It has shown inhibition of ZIKV by targeting PI3K/Akt pathway (Chan et al., 2018). During evaluation of multiple BSL-4 viruses by cellulr kinase inhibitors, OSU-03012 has also shown inhibition of replication in Lassa, Ebola, Marburg and Nipah viruses (Mohr et al., 2018). BIBX 1382 dihydrochoride is ErbB kinase inhibitor and has demonstrated inhibition of glycoprotein-dependent entry of Lassa and Ebola virus (Mohr et al., 2018). These compounds elaborate the virus-host interactions and role of host cell kinases as target for broad spectrum antiviral agents. Due to bioavailability issues and a dose-limiting increase of liver enzymes, the development of BIBX 1382 was interrupted after phase I trials (Dittrich et al., 2002).

ON108110 (ON-108110, ON 108110) is a novel small molecule, multi-kinase inhibitor of protein kinase CK2 and CDK-4/6. It has demonstrated inhibition of influenza A virus replication by restricting viral RNA synthesis (Figure 8) (Martinez-Gil et al., 2013). Similarly a novel antiviral compound, WV970 was identified through cell-based screening of a compound library followed by lead optimization (Figure 8) (Sasaki et al., 2014). The activity was observed in nanomolar range against influenza A and B viruses through inhibition of vRNP-mediated viral genome replication and transcription. The multiple kinase inhibitory properties as replication inhibitor were identified during structure-based and comprehensive human genome screenings and confirmed that several kinases are associated with viral replication as promising drug targets.

Drug Repurposing and Anti-Viral Kinase Inhibitors

For most fruitful outcome based drug discovery of a new drug it is necessary to start with an old drug (Sir James Black, 1998 Nobel Prize in Physiology and Medicine Laureate). In biomedical field, the scientific and technological advances in the last few decades have failed to witness significant success rate of the drug discovery process i.e., de novo and from the bench to the bedside (Strittmatter, 2014). Based on the clinical success rate, biopharmaceutical experts have estimated that only 12% of drug candidates entering into Phase I are successful (www.fda.gov; www.pharma.org). The causes of low success rate of classic drug discovery are target specificity, lack of experimental models to reproduce biological complexity, regulatory stringency, drug tolerability, and side effects. In addition to this the expenditure involved in the total drug discovery process is 1.5 – 2.6 billion US $ and time span of 13-15 years (Ashburn and Thor, 2004; Pushpakom et al., 2018). In such scenario, the drawbacks in the available drug discovery process can be resolved by application of drug repurposing strategies. Drug repurposing or repositioning involves a broad term and consists of different, experimental approaches to explore new therapeutic applications significantly diverse from the original medical indication (Pushpakom et al., 2018). In simplified terms, it is a process of identifying and validating new therapeutic indication for an existing or clinical/preclinical drug (Strittmatter, 2014; Pushpakom et al., 2018; Devillers, 2018). This approach offers advantage of bypassing long, risky and expensive preliminary drug discovery phases by use of available human clinical, pharmacokinetic and safety data which can be used as basis for drug development. 30% of the newly FDA approved drugs are identified by drug repurposing approach (Hernandez et al., 2017).

Drug repurposing processes can be classified as (i) Serendipitous observations, (ii) Target-based repurposing, (iii) Phenotypic screening, and (iv) *In silico*- assisted repurposing.

Serendipitous Observations

Serendipity i.e., "discovery of something not sought" has played a pronounced role in the discoveries of the earlier period. It is one of the factors contributing to drug discovery. Some of the well known examples are discovery of aniline purple, penicillin, lysergic acid diethylamide, meprobamate, chlorpromazine, imipramine, potassium bromide, chloral hydrate, and lithium (Thomas, 2006). Some of them have lead to drug repurposing based on the serendipitous observations recorded during preclinical, clinical or post-approval clinical trials. Further studies lead to subsequent rationalized evaluation and validation of the new treatment potential (Ashburn and Thor, 2004). An interesting example of drug repurposing is Thalidomide which was initially introduced as an antinausea drug for the treatment of morning sickness during pregnancy. It had teratogenicity and required to be removed from the market. Further studies on Thalidomide and its derivatives concluded it is used for repurposing in the treatment of leprosy and multiple myeloma (Rehman et al., 2011). Sildenafil is the next known drug from this category and was originally intended use in the treatment of hypertension. During clinical phase, the observed side-effects ended in its approval for use in erectile dysfunction and for the treatment of pulmonary hypertension (Ghofrani et al., 2006). The approach fails to explore full potential of drug repurposing as it is solely based on the casual observation and thus demands controlled and systematic methodologies.

Target-Based Repurposing

Target-based repurposing requires knowledge of the specific molecular drug target which is intended to be repurposed. The repurposing potential of the drug can be determined by carrying out further studies related to the role of target in a disease other than the original indication. The target may have significantly different roles in different diseases or pathological conditions. Favipiravir shows viral RNA polymerase inhibition in both influenza and Ebola viruses while imatinib, anti-cancer drug targets the Abelson tyrosine-protein kinase 2 (Abl2) required for viral replication in MERS-CoV and for efficient fusion and release of severe acute respiratory syndrome coronavirus

(SARS-CoV) (Coleman et al., 2016). In case of polypharmacology related features single drug acts on multiple targets of a unique disease pathway or a single drug acts on multiple targets of multiple disease pathways (Hopkins, 2008 and 2009; Reddy and Zhang, 2013). It is helpful in resolving undesirable adverse effects, side effects or toxicity which may lead to drug repurposing. Some remarkable studies have been reported for drugs initially having unique therapeutic target and later identified to have an average of 6 to 13 different targets (Medina-Franco et al., 2013; Naveja et al., 2016). The major limitation of the target-based drug repurposing model is its dependence on the existing scientific knowledge of the mechanism of action of the drug and other potential alternative targets.

Phenotypic Screening

The repurposing potential of a drug can be completely anticipated only on availability of its molecular targets and their characterization details. In phenotypic screening of bioactive molecule libraries in different experimental cell based or *in vivo* disease models is carried out without the requirement of prior knowledge of their target and/or mechanism of action. The highthroughput phenotype screening has contributed to drug discovery but it may result in number of false positive hits (Swinney and Anthony, 2011; Bastos and Coelho, 2014). For phenotypic screening, various small molecules libraries can be used to accelerate drug repurposing (Swinney and Anthony, 2011; Loregian and Palu, 2013; Bastos and Coelho, 2014; Jones and Bunnage, 2017).

In Silico-Assisted Repurposing

The *in silico* or computer-based approaches have improved acceptance in the field of drug discovery with the advent of big data and systems biology. The understanding of mechanism of action of drug is approachable with the availability of transcriptomic, proteomic, metabolomic data and in addition to this the *in silico* data mining and modeling tools have pushed our capacity to analyze data to the next level (Knox et al., 2011; Hurle et al., 2013; Ferrero et al., 2017). Some of the *in silico* methods applied in drug discovery and drug repurposing include the screening of chemical, biological, and text databases, analysis of quantitative structure-activity relationships, pharmacophores, homology models, other molecular modeling approaches, network analysis of biological functions, machine learning and almost any other analysis tools that include use of a computer (Ekins et al., 2007; Tari and Patel, 2014; Cichonska et al., 2015; Sidders et al., 2018). The data mining approaches for biological, chemical and clinical datasets, has proved effective in unveiling novel relationships (Chatterjee et al., 2009; Oprea et al., 2011) and for further information of indication identification can be added by combining the available databases with *in vitro* experimental molecule screenings (Chiiang and Butte, 2009; Platz et al., 2011). The computer-assisted drug repurposing requires use of integrated use of the predictive and analytic capacity of *in silico* tools with some of the target biased or unbiased experimental evaluation/validation methods previously mentioned. This "systems pharmacology" approach (van der Graaf et al., 2011; Allarakhia, 2013; Zhou et al., 2016) is informative to design more comprehensive repurposing strategies with more effective predictive capacity and, hopefully, improved candidate success rates.

DRUG REPURPOSING OF KINASE INHIBITORS

The evolving nature of viruses, high transmissibility, host promiscuity, subtopical vaccine efficacy, limited antiviral arsenal, and zoonotic, and pandemic potential of various viruses make them an attractive target to adopt drug repurposing strategies during antiviral drug discovery. The reported drug repurposing strategies have been applied to screen specific drug libraries of kinase inhibitors but failed to give a successful drug candidate (Table 4) (Ludwig, 2014; Perwitasari et al., 2015). Some of the drugs which entered in preclinical and clinical phases of development are LASAG (BAY81-8781), Trametinib and Ruxolitinib.

Table 4. Repurposed kinase inhibitors for the treatment of influenza viral infections

Drug	Initial indication	Initial activity	Anti-influenza activity	Status	Ref
LASAG (BAY81-8781)	Anti-inflammatory	NF-kB inhibitor	NF-kB inhibition	Phase II (2012-004072-19)	Droebner et al., 2017; Scheuch et al., 2018
Trametinib	Approved anti-cancer drug	MEK1/2 inhibitor	vRNP transport inhibition	Preclinical	Schräder et al., 2018
Ruxolitinib	Approved for myelofibrosis treatment	JAK inhibitor	Virion formation & vRNA incorporation inhibition	Preclinical	Watanabe et al., 2014

LASAG (BAY81-87981)

LASAG (BAY81-87981) (D, L-Lysine acetylsalicylate-glycine), is a derivative of anti-inflammatory drug acetylsalicylic acid (ASA). It has been identified as an anti-influenza drug through drug repurposing strategies i.e., serendipitous and target-based. ASA has shown *in vitro* and *in vivo* anti-influenza effects via NF-kB inhibition. The NF-kB activating kinase IkkB, negatively impacts influenza vRNP transport and release of infectious viral particles (Yin et al., 1998; Wurzer et al., 2004; Mazur et al., 2007). The pharmacokinetic limitations of ASA demanded the development of improved derivative i.e., LASAG with improved stability and tolerability. LASAG demonstrated *in vitro* antiviral activity against several human and avian influenza viruses. In a mouse infection model, inhalation of LASAG resulted in reduced lung viral titers and protection of mice from lethal infection (Droebner et al., 2017). During Phase II studies, aerosolized LASAG versus placebo for treatment of severe influenza (all patients receiving Tamiflu as standard of care treatment) demonstrated improved time for symptom alleviation compared to placebo, despite the absence of a statistically significant reduction of viral load in LASAG-treated group (Scheuch et al., 2018).

Trametinib

In 2013, FDA approved trametinib (GSK-1120212), a mitogen-activated extracellular signal-regulated kinase (MEK1/2) inhibitor for the treatment of melanomas (Lugowska et al., 2015; Lin et al., 2018). Trametinib is highly selective with side effect and has been selected for a repurposing approach to reduce the anti-flu drug development time. During *in vitro* and *in vivo* studies, Trametinib efficiently blocks replication of different influenza A virus (IAV) subtypes and has ability to interfere with export of progeny vRNPs from the nucleus. It possesses broad antiviral activity against various IAV strains and limited hyper-expression of several cytokines. Thus, we show for the first time that a clinically approved MEK inhibitor acts as a potent anti-influenza agent (Schräder et al., 2018).

Ruxolitinib

Ruxolitinib became the first approved therapy for myelofibrosis in United States and Europe based on the two large phase III trials (COMFORT studies) (Figure 9) (Harrison et al., 2012). It is potent and selective inhibitor of Janus kinase (JAK1/2) signal transducer and STAT pathway essential for host immunity and defense. Watanabe et al., (2014) have reported systematic analysis of various host factors required for influenza virus proteins and replication. Based on the identified 1,000 host factors and the stage of the viral life cycle affected, a virus-host protein interaction network was generated. Inhibitors of these host factors including ruxolitinib were used to validate several antiviral drug targets i.e.,Golgi-specific brefeldin A-resistant guanine nucleotide exchange factor 1 (GBF1) and JAK1 (Watanabe et al., 2014).Additionally, ruxolitinib has shown inhibition of influenza A virus mediated changes in neutrophil gene expression (Mortimer et al., 2018).

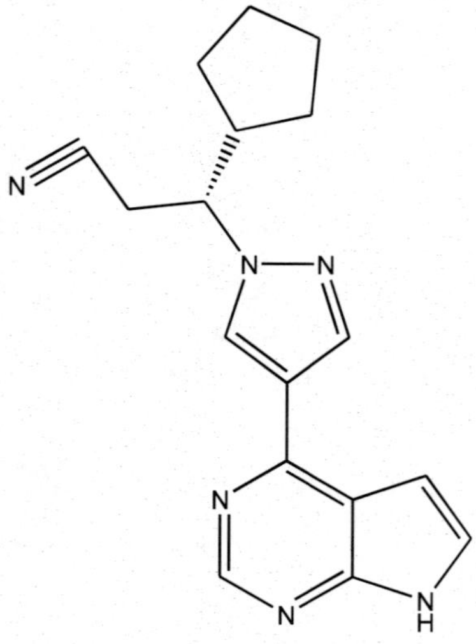

Figure 9. Chemical structure of Ruxolitinib.

Urea-Based Kinase Inhibitors (UBKIs)

The approved urea-based kinase inhibitors (UBKIs), regorafenib and sorafenib possess anti-influenza virus activity with low cytotoxicity. UBKIs shows antiviral activity by blocking IV replication after internalization and were found active against cowpox virus (CPXV), hantavirus (HTV), herpes simplex virus 1 (HSV1) and vesicular stomatitis virus (VSV) (Lesch et al., 2019). The broad-spectrum of antiviral activity of the selected UBKIs and substantial robustness against resistance development makes them attractive host-directed drugs against influenza and other viral infections (Lesch et al., 2019).

Conclusion

Lack of therapeutic strategies for most viral infections demands major medical attention. To develop effective therapeutics the understanding of molecular mechanisms of viral replication and pathogenesis, virus-host interactions is a major challenge against present and future emerging viruses. The role of host-cell kinases in viral life cycle is always an attractive target for development of newer antiviral drugs with well established mechanism of action. Viral protein kinases constitute important virulence factor and are key elements in various viruses. They have supported the viruses during host evolution to maintain their ability to establish lifelong infections by adapting the improved complexity of their host. The viral protein kinases are multifunctional (phosphorylation, alteration of viral and cellular targets, replication, etc.) and lack of viral kinase activity has shown reduced virulence. The viral protein kinases have been evaluated as possible antiviral drug target using compounds from different chemical categories (Maribavir, Cidofovir, BisI). Various kinase inhibitors have been successfully developed as antiviral agents. In addition to this an effective approach to identify and develop ready-to-use antiviral agents is drug repurposing. Additional benefits of this approach are identification of new pathways, viral targets and to skip preclinical and phase I (safety/tolerability) clinical trials. The identified drug directly enters phase 2 clinical trials (for efficacy testing) and thus accelerated the drug discovery and development process. The success observed with drug repurposing approach to search new targets for therapeutic interventions is encouraging.

Acknowledgment

Work by the authors has been supported by the facilities provided by KIET Group of Institutions, GZB, and Lala Lajpat Rai Memorial Medical College, Meerut, India.

References

Alessi, D. R., James, S. R., Downes, C. P., Holmes, A. B., Gaffney, P. R. J., Reese, C. B. & Cohen, P. (1997). Characterization of a 3-

phosphoinositide-dependent protein kinase which phosphorylates and activates protein kinase B[alpha]. *Curr Biol*, *7*, 261–269.

Alessi, D. R., Kozlowski, M. T., Weng, Q. P., Morrice, N. & Avruch, J. (1998). 3-Phosphoinositide-dependent protein kinase 1 (PDK1) phosphorylates and activates the p70 S6 kinase *in vivo* and *in vitro*. *Curr Biol*, *8*, 69–81.

Allarakhia, M. (2013). Open-source approaches for the repurposing of existing or failed candidate drugs: learning from and applying the lessons across diseases. *Drug Des Devel Ther*, *7*, 753–766.

Aoki, T., Hirono, I., Kurokawa, K., Fukuda, H., Nahary, R., Eldar, A., Davison, A. J., Waltzek, T. B., Bercovier, H. & Hedrick, R. P. (2007). Genome sequences of three koi herpesvirus isolates representing the expanding distribution of an emerging disease threatening koi and common carp worldwide. *J Virol*, *81*, 5058–5065.

Arthur, L. O., Bess, J. W., Sowder, R. C., Benveniste, R. E., Mann, D. L., Chermann, J.C. & Henderson, L. E. (1992). Cellular proteins bound to immunodeficiency viruses: implications for pathogenesis and vaccines. *Science*, *258*, 1935-1938.

Ashburn, T. T. & Thor, K. B. (2004). Drug repositioning: identifying and developing new uses for existing drugs. *Nat Rev Drug Discov*, *3*, 673–683.

Badia, R., Angulo, G., Riveira-Munoz, E., Pujantell, M., Puig, T., Ramirez, C., Torres-Torronteras, J., Marti, R., Pauls, E., Clotet, B., Ballana, E. & Este, J. A. (2016). Inhibition of herpes simplex virus type 1 by the CDK6 inhibitor PD-0332991 (palbociclib) through the control of SAMHD1. *J Antimicrob Chemother*, *71*, 387–394.

Balendran, A., Hare, G. R., Kieloch, A., Williams, M. R. & Alessi, D. R. (2000). Further evidence that 3-phosphoinositide-dependent protein kinase-1 (PDK1) is required for the stability and phosphorylation of protein kinase C (PKC) isoforms. *FEBS Lett*, *484*, 217–223.

Bastos, L. F. S. & Coelho, M. M. (2014). Drug repositioning: playing dirty to kill pain. *CNSDrugs*, *28*, 45–61.

Bekerman, E., Neveu, G., Shulla, A., Brannan, J., Pu, S.-Y., Wang, S., Xiao, F., Barouch-Bentov, R., Bakken, R. R., Mateo, R., Govero, J.,

Nagamine, C. M., Diamond, M. S., De Jonghe, S., Herdewijn, P., Dye, J. M., Randall, G. & Einav, S. (2017). Anticancer kinase inhibitors impair intracellular viral trafficking and exert broad-spectrum antiviral effects. *J Clin Invest*, *127*, 1338–1352.

Bhattacharya, D., Ansari, I. H. & Striker, R. (2009a). The flaviviral methyltransferase is a substrate of Casein Kinase 1. *Virus Res*, *141*(1), 101–104.

Bhattacharya, D., Mayuri, Best S. M., Perera, R., Kuhn, R. J. & Striker, R. (2009b). Protein kinase G phosphorylates mosquito-borne flavivirus NS5. *J Virol*, *83*(18), 9195–9205.

Biron, K. K., Harvey, R. J., Chamberlain, S. C., Good, S. S., Smith, A. A. 3rd, Davis, M. G., Talarico, C. L., Miller, W. H., Ferris, R., Dornsife, R. E., Stanat, S. C., Drach, J. C., Townsend, L. B. & Koszalka, G. W. (2002). Potent and selective inhibition of human cytomegalovirus replication by 1263W94, a benzimidazole L-riboside with a unique mode of action. *Antimicrob Agents Chemother*, *46*, 2365–2372.

Briant, L., Reynes, J., Coudronniere, N., Benezech, J. P. & Devaux, C. (1999). HIV-1 reactivation in resting peripheral blood mononuclear cells of infected adults upon *in vitro* CD4 cross-linking by ligands of the CDR2-loop in extracellular domain 1. *J Acquir Immune Defic Syndr*, *21*, 9-19.

Buchkovich, N. J., Yu, Y., Zampieri, C. A. & Alwine, J. C. (2008). The TORrid affairs of viruses: effects of mammalian DNA viruses on the PI3K–Akt–mTOR signalling pathway. *Nat Rev Microbiol*, *6*, 266–275.

Bukong, T. N., Hall, W. W. & Jacque, J. M. (2010). Lentivirus-associated MAPK/ERK2phosphorylates EMD and regulates infectivity. *J Gen Virol*, *91*, 2381-2392.

Cartier, C., Deckert, M., Grangeasse, C., Trauger, R., Jensen, F., Bernard, A., Cozzone, A., Desgranges, C. & Boyer, V. (1997). Association of ERK2 Mitogen-Activated Protein Kinase with human immunodeficiency virus particles. *J Virol*, *71*, 4832-4837.

Cartier, C., Hemonnot, B., Gay, B., Bardy, M., Sanchiz, C., Devaux, C. & Briant, L. (2003). Active cAMP-dependent protein kinase incorporated within highly purified HIV-1 particles is required for viral infectivity and interacts with viral capsid protein. *J Biol Chem*, *278*, 35211-35219.

Cen, L., Hsieh, F.C., Lin, H.J., Chen, C.S., Qualman, S. J. & Lin, J. (2007). PDK-1/AKT pathway as a novel therapeutic target in rhabdomyosarcoma cells using OSU-03012 compound. *British J Cancer*, *97*, 785-791.

Chan, J. F. W., Zhu, Z.,Chu, H., Yuan, S.,Chik, K. K. H., Chan, C. C. S., Poon, V. K. M., Yip, C. C. Y., Zhang, X.,Tsang, J. O. L., Zou, Z., Tee, K. M. T., Shuai, H., GangLu, G. &Yuen, K. Y. (2018). The celecoxib derivative kinase inhibitor AR-12 (OSU-03012) inhibits Zika virus via down-regulation of the PI3K/Akt pathway and protects Zika virus-infected A129 mice: A host-targeting treatment strategy. *Antivir Res*, *160*, 38-47.

Chart Pack: Biopharmaceuticals in Perspective, Summer 2018. *Phrma* (2018). Available online at: http://www.phrma.org/report/ chart-packbiopharmaceuticals-in-perspective-summer-2018 (Accessed September 28, 2019).

Chatterjee, S., Szustakowski, J. D., Nanguneri, N. R., Mickanin, C., Labow, M. A., Nohturfft, A., Dev, K. K. & Sivasankaran, R. (2009). Identification of novel genes and pathways regulating SREBP transcriptional activity. *PLoS ONE*, *4*, 5197.

Cheng, P.H., Rao, X.M., McMasters, K. M. & Zhou, H. S. (2013). Molecular basis for viral selective replication in cancer cells: activation of CDK2 by adenovirus-induced Cyclin E. *PLoS One*, 8, e57340.

Chertova, E., Chertov, O., Coren, L. V., Roser, J. D., Trubey, C. M., Bess, J. W. Jr., Sowder, R. C., Barsov, E., Hood, B. L., Fisher, R. J.,Nagashima, K., Conrads, T. P., Veenstra, T. D., Lifson, J. D. & Ott, D. E. (2006). Proteomic and biochemical analysis of purified human immunodeficiency virus type 1 produced from infected monocyte-derived macrophages. *J Virol*, *80*, 9039-9052.

Chiang, A. P. & Butte, A. J. (2009). Systematic evaluation of drug-disease relationships to identify leads for novel drug uses. *Clin Pharmacol Ther*, *86*, 507–510.

Chong, Z. Z., Shang, Y. C., Wang, S. & Maiese, K. (2012). A critical kinase cascade in neurological disorders: PI3K, Akt and mTOR. Fiuture *Neurol*, *7*, 733–748.

Chou, S., Van Wechel, L. C. & Marousek, G. I. (2006). Effect of cell culture conditions on the anticytomegalovirus activity of maribavir. *Antimicrob Agents Chemother*, *50*, 2557–2559.

Chu, J. J. & Yang, P.L. (2007). c-Src protein kinase inhibitors block assembly and maturation of dengue virus. *Proc Natl Acad Sci U S A*, *104*, 3520–3525.

Cichonska, A., Rousu, J. & Aittokallio, T. (2015). Identification of drug candidates and repurposing opportunities through compound-target interaction networks. *Expert Opin Drug Discov*, *10*, 1333–1345.

Clark, M. J., Miduturu, C., Schmidt, A. G., Zhu, X., Pitts, J. D., Wang, J., Potisopon, S., Zhang, J., Wojciechowski, A., Hann Chu, J. J., Gray, N. S. & Yang, . L. (2016). GNF-2 inhibits dengue virus by targeting Abl kinases and the viral E protein. *Cell Chem Biol*, *23*, 443–452.

Coleman, C. M., Sisk, J. M., Mingo, R. M., Nelson, E. A., White, J. M. & Frieman, M. B. (2016). Abelson kinase inhibitors are potent inhibitors of severe acute respiratory syndrome coronavirus and middle east respiratory syndrome coronavirus fusion. *J Virol*, *90*, 8924–8933.

Collett, M. S. & Erikson. R. L. (1978). Protein kinase activity associated with the avian sarcoma virus src gene product. *Proc Natl Acad Sci U S A*, *75*, 2021–2024.

Coussens, L., Parker, P. J., Rhee, L., Yang-Feng, T. L., Chen, E., Waterfield, M. D., Francke, U. & Ullrich, A. (1986). Multiple, distinct forms of bovine and human protein kinase C suggest diversity in cellular signaling pathways. *Science*, *233*, 859–867.

Coyne, C. B., & Bergelson, J. M. (2006). Virus-induced Abl and Fyn kinase signals permit coxsackievirus entry through epithelial tight junctions. *Cell*, *124*, 119–131.

de Wispelaere, M., LaCroix, A. J. & Yang, P. L. (2013). The small molecules AZD0530 and dasatinib inhibit dengue virus RNA replication via Fyn kinase. *J Virol*, *87*, 7367–7381.

Devaux, C., Briant, L. & Biard-Piechaczyk, M. (1999). Involvement of CD4 and CXCR4 molecules in HIV-1 gp120-induced T cell activation and apoptosis. In *Recent Research Developments in Virology*" (S. Pandalai, Ed.), Vol. 1, pp. 139-155. Transworld Research Network, Trivandrum.

Devaux, C., Boucraut, J., Poirier, G., Corbeau, P., Rey, F., Benkirane, M., Perarnau, B., Kourilsky, F. & Chermann, J. C. (1990). Anti-beta 2-microglobulin monoclonal antibodies mediate a delay in HIV1 cytopathic effect on MT4 cells. *Res Immunol*, *141*, 357-372.

Devillers, J. (2018). Repurposing drugs for use against Zika virus infection. *SAR QSAR Environ Res*, *29*, 103–115.

Diao, J., Pantua, H., Ngu, H., Komuves, L., Diehl, L., Schaefer, G. & Kapadia, S. B. (2012). Hepatitis C virus induces epidermal growth factor receptor activation via CD81 binding for viral internalization and entry. *J Virol*, *86*, 10935–10949.

Dittrich, C., Greim, G., Borner, M., Weigang-Kohler, K., Huisman, H., Amelsberg, A., Ehret, A., Wanders, J., Hanauske, A. & Fumoleau, P. (2002). Phase I and pharmacokinetic study of BIBX 1382 BS, an epidermal growth factor receptor (EGFR) inhibitor, given in a continuous daily oral administration. *Eur J Cancer*, *38*, 1072-1080.

Droebner, K., Haasbach, E., Dudek, S. E., Scheuch, G., Nocker, K., Canisius, S., Ehrhardt, C., von Degenfeld, G., Ludwig, S. & Planz, O. (2017). Pharmacodynamics, pharmacokinetics, and antiviral activity of BAY 81- 8781, a novel NF-κB inhibiting anti-influenza drug. *Front Microbiol*, *8*, 2130.

Drug Innovation. *Novel Drug Approvals for 2018*. (2018). Available online at: https://www.fda.gov/Drugs/DevelopmentApprovalProcess/DrugInnovation/ucm592464.htm (Accessed August 18, 2019).

Dyall, J., Coleman, C. M., Hart, B. J., Venkataraman, T., Holbrook, M. R., Kindrachuk, J., Johnson, R. F., Olinger, G. G., Jr., Jahrling, P. B., Laidlaw, M., Johansen, L. M., Lear-Rooney, C. M., Glass, P. J., Hensley, L. E. & Frieman, M. B. (2014). Repurposing of clinically developed drugs for treatment of Middle East respiratory syndrome coronavirus infection. *Antimicrob Agents Chemother*, *58*, 4885–4893.

Ehrhardt, C., Marjuki, H., Wolff, T., Nurnberg, B., Planz, O., Pleschka, S. & Ludwig, S. (2006). Bivalent role of the phosphatidylinositol-3-kinase (PI3K) during influenza virus infection and host cell defence. *Cell Microbiol*, *8*, 1336–1348.

Eierhoff, T., Hrincius, E. R., Rescher, U., Ludwig, S. & Ehrhardt, C. (2010). The Epidermal Growth Factor Receptor (EGFR) Promotes uptake of Influenza A Viruses (IAV) into host cells. *PLoS Pathog*, *6*, e1001099.

Ekins, S., Mestres, J. & Testa, B. (2007). In silico pharmacology for drug discovery: methods for virtual ligand screening and profiling. *Br J Pharmacol*, *152*, 9–20.

Fabbro, D., Cowan-Jacob, S. W. & Moebitz, H. (2015). Ten things you should know about protein kinases: IUPHAR review 14. *Br J Pharmacol*, *172*, 2675–2700.

Ferrero, E., Dunham, I. & Sanseau, P. (2017). In silico prediction of novel therapeutic targets using gene–disease association data. *J Transl Med*, *15*, 182.

Garcia, M., Cooper, A., Shi, W., Bornmann, W., Carrion, R., Kalman, D. & Nabel, G. J. (2012). Productive replication of Ebola virus is regulated by the c-Abl1 tyrosine kinase. *Sci Transl Med*, *4*, 123-124.

Ghofrani, H. A., Osterloh, I. H. & Grimminger, F. (2006). Sildenafil: from angina to erectile dysfunction to pulmonary hypertension and beyond. *Nat Rev Drug Discov*, *5*, 689–702.

Gill, R. B., Frederick, S. L., Hartline, C. B., Chou, S. & Prichard, M. N. (2009). Conserved retinoblastoma protein-binding motif in human cytomegalovirus UL97 kinase minimally impacts viral replication but affects susceptibility to maribavir. *Virol J*, *6*, 9.

Giroud, C., Chazal, N. & Briant, L. (2011). Cellular kinases incorporated into HIV-1 particles: passive or active passengers? *Retrovirology*, *8*, Article Number 71.

Graf, L., Webel, R., Wagner, S., Hamilton, S. T., Rawlinson, W. D., Sticht, H. &Marschall, M. (2013). The cyclin-dependent kinase ortholog pUL97 of human cytomegalovirus interacts with cyclins. *Viruses*, *5*(12), 3213-3230.

Guendel, I., Agbottah, E. T., Kehn-Hall, K. & Kashanchi, F. (2010). Inhibition of human immunodeficiency virus type-1 by cdk inhibitors. *AIDS Res Ther*, *7*, 7.

Habran, L., Bontems, S., Di Valentin, E., Sadzot-Delvaux, C. & Piette, J. (2005). Varicella-zoster virus IE63 protein phosphorylation by roscovitine-sensitive cyclin-dependent kinases modulates its cellular localization and activity. *J Biol Chem*, *280*, 29135–29143.

Hale, B. G., Jackson, D., Chen, Y. H., Lamb, R. A. & Randall, R. E. (2006). Influenza A virus NS1 protein binds p85beta and activates phosphatidylinositol-3-kinase signaling. *Proc Natl Acad Sci U S A*, *103*, 14194–14199.

Hanks, S. K. & Hunter, T. (1995). Protein kinases 6: the eukaryotic protein kinase superfamily: kinase (catalytic) domain structure and classification. *FASEBJ*, *9*, 576–596.

Hansen, J. E., Nielsen, C., Mathiesen, L. R. & Nielsen, J. O. (1991). Involvement of lymphocyte function-associated antigen-1 (LFA-1) in HIV infection: inhibition by monoclonal antibody. *Scand J Infect Dis*, *23*, 31-36.

Harrison, C. & Vannucchi, A. M. (2012). Ruxolitinib: a potent and selective Janus kinase 1 and 2 inhibitor in patients with myelofibrosis. An update for clinicians. *Ther Adv Hematol*, *3*(6), 341–354.

Hatanaka, M., Twiddy, E. & Gilden, R. V. (1972). Protein kinase associated with RNA tumour viruses and other budding RNA viruses. *Virology*, *47*, 536-538.

Hemonnot, B., Cartier, C., Gay, B., Rebuffat, S., Bardy, M., Devaux, C., Boyer, V. & Briant, L. (2004). The host cell MAP kinase ERK-2 regulates viral assembly and release by phosphorylating the p6gag protein of HIV-1. *J Biol Chem*, *279*, 32426-32434.

Hemonnot, B., Molle, D., Bardy, M., Gay, B., Laune, D., Devaux, C. & Briant, L. (2006). Phosphorylation of the HTLV-1 matrix L-domain-containing protein by virus-associated ERK-2 kinase. *Virology*, *349*, 430-439.

Herget, T., Freitag, M., Morbitzer, M., Kupfer, R., Stamminger, T. & Marschall, M. (2004). Novel chemical class of pUL97 protein kinase-specific inhibitors with strong anticytomegaloviral activity. *Antimicrob Agents Chemother*, *48*(11), 4154-4162.

Hernandez, J. J., Pryszlak, M., Smith, L., Yanchus, C., Kurji, N., Shahani, V. M. & Molinski, S. V. (2017). Giving drugs a second chance: overcoming regulatory and financial hurdles in repurposing approved drugs as cancer therapeutics. *Front Oncol*, *7*, 273.

Hirsch, A. J., Medigeshi, G. R., Meyers, H. L., DeFilippis, V., Fru ̈h, K., Briese, T., Lipkin, W. I. & Nelson, J. A. (2005). The Src family kinase c-yes is required for maturation of West Nile virus particles. *J Virol*, *79*, 11943–11951.

Hopkins, A. L. (2008). Network pharmacology: the next paradigm in drug discovery. *Nat Chem Biol*, *4*, 682–690.

Hopkins, A. L. (2009). Drug discovery: predicting promiscuity. *Nature*, *462*, 167–168.

Houts, G. E., Miyagi, M., Ellis, C., Beard, D., Watson, K.F. & Beard, J. W. (1978). Protein kinasefrom avian myeloblastosis virus. *J Virol*, *25*, 546-552.

Hume, A. J., Finkel, J. S., Kamil, J. P., Coen, D. M., Culbertson, M. R. & Kalejta, R. F. (2008). Phosphorylation of retinoblastoma protein by viral protein with cyclin-dependent kinase function. *Science*, *320*, 797–799.

Hurle, M. R., Yang, L., Xie, Q., Rajpal, D. K., Sanseau, P. & Agarwal, P. (2013). Computational drug repositioning: from data to therapeutics. *Clin Pharmacol Ther*, *93*, 335–341.

Hutterer, C., Eickhoff, J., Milbradt, J., Korn, K., Zeitträger, I., Bahsi, H., Wagner, S., Zischinsky, G., Wolf, A., Degenhart, C., Unger, A., Baumann, M., Klebl, B. & Marschall, M. (2015). A Novel CDK7 Inhibitor of the Pyrazolotriazine Class Exerts Broad-Spectrum Antiviral Activity at Nanomolar Concentrations. *Antimicrob Agents Chemother*, *59*(4), 2062–2071.

Jacob, T., Van den Broeke, C. & Favoreel, H. W. (2011). Viral serine/threonine protein kinases. *J Virol*, *85*, 1158-1173.

Jacque, J. M., Mann, A., Enslen, H., Sharova, N., Brichacek, B., Davis, R. J. & Stevenson, M. (1998). Modulation of HIV-1 infectivity by MAPK, a virion associated kinase. *EMBOJ*, 17, 2607-2618.

Jakob, N. J., Müller, K., Bahr, U. & Darai, G. (2001). Analysis of the first complete DNA sequence of an invertebrate iridovirus: coding strategy of the genome of Chilo iridescent virus. *Virology*, *286*, 182–196.

Jaw, J., Hill, P. & Goodman, D. (2017). Combination of leflunomide and everolimus for treatment of BK virus nephropathy. *Nephrology*, *22*, 326–329.

Johnson, J. C., Martinez, O., Honko, A. N., Hensley, L. E., Olinger, G. G. & Basler, C. F. (2014). Pyridinyl imidazole inhibitors of p38 MAP kinase impair viral entry and reduce cytokine induction by Zaire ebolavirus in human dendritic cells. *Antiviral Res*, *107*, 102–109.

Jones, L. H. & Bunnage, M. E. (2017). Applications of chemogenomic library screening in drug discovery. *Nat Rev Drug Discov*, *16*, 285–296.

Jung, Y. H., Moon, K. C., Ha, J. W., Kim, S.-J., Ha, I.-S., Cheong, H. I. & Kang, H. G. (2013). Leflunomide therapy for BK virus allograft nephropathy after pediatric kidney transplantation. *Pediatr Transplant*, *17*, E50–E54.

Kalejta, R. F. (2008). Tegument proteins of human cytomegalovirus. *Microbiol Mol Biol Rev*, *72*, 249–265.

Kamil, J. H., Hume, A. J., Jurak, I., Münger, K., Kalejta, R. F. & Coen, D. M. (2009). Human papillomavirus 16 E7 inactivator of retinoblastoma family proteins complements human cytomegalovirus lacking UL97 protein kinase. *Proc Natl Acad Sci U S A*, *106*, 16823–16828.

Kindrachuk, J., Ork, B., Hart, B. J., Mazur, S., Holbrook, M. R., Frieman, M. B., Traynor, D., Johnson, R. F., Dyall, J., Kuhn, J. H., Olinger, G. G., Hensley, L. E. & Jahrling, P. B. (2015). Antiviral potential of ERK/MAPK and PI3K/AKT/ mTOR signaling modulation for middle east respiratory syndrome coronavirus infection as identified by temporal kinome analysis. *Antimicrob Agents Chemother*, *59*, 1088–1099.

Kittler, H. & Tschand, P. (2018). Driver mutations in the mitogen-activated protein kinase pathway: the seeds of good and evil. *Br J Dermat*, *178*, 26-27.

Knox, C., Law, V., Jewison, T., Liu, P., Ly, S., Frolkis, A., Pon, A.,Banco, K., Mak, C., Neveu, V., Djoumbou, Y., Eisner, R., Guo, A. C. &

Wishart, D. S. (2011). DrugBank 3.0: a comprehensive resource for "omics" research on drugs. *Nucleic Acids Res*, *39*, D1035–1041.

Kolokoltsov, A. A., Saeed, M. F., Freiberg, A. N., Holbrook, M. R. & Davey, R. A. (2009). Identification of novel cellular targets for therapeutic intervention against Ebola virus infection by siRNA screening. *Drug Dev Res*, *70*, 255–265.

Köstler, W. J. & Zielinski, C. C. (2015). Targeting Receptor Tyrosine Kinases in Cancer. In *Receptor Tyrosine Kinases: Structure, Functions and Role in Human Disease*. New York: Spring; 2015. p. 225–278.

Kumar, R., Agrawal, T., Khan, N. A., Nakayama, Y. & Medigeshi, G. R. (2016). Identification and characterization of the role of c-terminal Src kinase in dengue virus replication. *SciRep*, *6*, 30490.

Kuny, C. V., Chinchilla, K., Culbertson, M. R. & Kalejta, R. F. (2010). Cyclin-dependent kinase-like function is shared by the beta and gamma subsets of the conserved herpesvirus protein kinases. *PLoSPathog*, *6*, e1001092.10.1371.

Langhammer, S., Koban, R., Yue, C. & Ellerbrok, H. (2011). Inhibition of poxvirus spreading by the anti-tumor drug Gefitinib (Iressa). *AntiviralRes*, *89*, 64–70.

Le Good, J. A., Ziegler, W. H., Parekh, D. B., Alessi, D. R., Cohen, P. & Parker, P. J. (1998). Protein kinase C isotypes controlled by phosphoinositide 3-kinase through the protein kinase PDK1. *Science*, *281*, 2042–2045.

Leader, D. P. & Katan, M. (1988). Viral aspects of protein phosphorylation. *J Gen Virol*, *69*, 1441–1464.

Leader, D. P. (1993). Viral protein kinases and protein phosphatases. *Pharmacol Ther*, *59*, 343–389.

Lesch, M., Luckner, M., Meyer, M., Weege, F., Gravenstein, I., Raftery, M., Sieben, C., Martin-Sancho, L., Imai-Matsushima, A., Welke, R.-W., Frise, R., Barclay, W., Schonrich, G., Herrmann, A., Meyer, T. F., Karlas, A. (2019). RNAi-based small molecule repositioning reveals clinically approved urea-based kinase inhibitors as broadly active antivirals. *PLoS Pathog*, *15*(3), e1007601.

Li, J., Li, H. & Tsai, M. D. (2003). Direct binding of the Nterminus of HTLV-1 tax oncoprotein to cyclin-dependent kinase 4 is a dominant path to stimulate the kinase activity. *Biochemistry*, *42*, 6921–6928.

Liacini, A., Seamone, M. E., Muruve, D. A. & Tibbles, L. A. (2010). Anti-BK virus mechanisms of sirolimus and leflunomide alone and in combination: toward a new therapy for BK virus infection. *Transplantation*, *90*, 1450–1457.

Lin, L., Dapeng Ding, D., Yanmei Jiang, Y., Yan Li, Y. &Li, S. (2018). MEK inhibitors induce apoptosis via FoxO3a-dependent PUMA induction in colorectal cancer cells. *Oncogenesis*, *7*, Article number: 67.

Littler, E., Stuart, A. D. & Chee, M. S. (1992). Human cytomegalovirus UL97 open reading frame encodes a protein that phosphorylates the antiviral nucleoside analogue ganciclovir. *Nature*, *358*, 160–162.

Loregian, A. & Palù, G. (2015). How academic labs can approach the drug discovery process as a way to synergize with big pharma. *Trends Microbiol*, *21*, 261–264.

Ludwig, S. (2014). Will omics help to cure the flu? *Trends Microbiol*, *22*, 232–233.

Lugowska, I., Koseła-Paterczyk, H., Kozak, K. & Rutkowski, P. (2015). Trametinib: a MEK inhibitor for management of metastatic melanoma. *Onco Targets Ther*, *8*, 2251–2259.

Lupberger, J., Zeisel, M., Xiao, F., Thumann, C., Fofana, I., Zona, L., Davis, C., Mee, C., Turek, M., Gorke, S., Royer, C., Fischer, B., Zahid, M. N., Lavillette. D., Fresquet, J., Cosset, F. L., Rothenberg, S. M., Pietschmann, T., Patel, A. H., Pessaux, P., Doffoël, M., Raffelsberger, W., Poch, O., McKeating, J. A., Brino, L. & Baumert, T. F. (2011). EGFR and EphA2 are host factors for hepatitis C virus entry and possible targets for antiviral therapy. *Nat Med*, *17*, 589–595.

Maeda, N., Fan, H. & Yoshikai, Y. (2008). Oncogenesis by retroviruses: old and new paradigms. *Rev Med Virol*, *18*, 387–405.

Malachowa, N., Freedman, B., Sturdevant, D. E., Kobayashi, S. D., Nair, V., Feldmann, F., Starr, T., Steele-Mortimer, O., Kash, J. C., Taubenberger, J. K., Feldmann, H. & DeLeo F. R. (2018). Differential

Ability of Pandemic and Seasonal H1N1 Influenza A Viruses To Alter the Function of Human Neutrophils. *mSphere*, *3* (1), e00567-17.

Manning, G., Whyte, D. B., Martinez, R., Hunter, T. & Sudarsanam, S. (2002). The protein kinase complement of the human genome. *Science*, 298, 1912–1934.

Manning, G., Whyte, D. B., Martinez, R., Hunter, T. & Sudarsanam, S. (2002). The protein kinase complement of the human genome. *Science*, *298*, 1912–1934.

Mannova´, P. & Beretta, L. (2005). Activation of the N-Ras– PI3K–Akt-mTOR pathway by hepatitis C virus: control of cell survival and viral replication. *J Virol*, *79*, 8742–8749.

Marschall M, Feichtinger S. & Milbradt, J. (2011). Regulatory roles of protein kinases in cytomegalovirus replication. *Adv Virus Res*, *80*, 69–101.

Marschall, M., Marzi, A., aus dem Siepen, P., Jochmann, R., Kalmer, M., Auerochs, S., Lischka, P., Leis, M. & Stamminger, T. (2005). Cellular p32 recruits cytomegalovirus kinase pul97 to redistribute the nuclear lamina. *J Biol Chem*, *280*, 33357–33367.

Martinez-Gil, L., Alamares-Sapuay, J. G., Ramana Reddy, M. V., Goff, P. H., Premkumar Reddy, E. & Palese, P. (2013). A small molecule multi-kinase inhibitor reduces influenza A virus replication by restricting viral RNA synthesis. *Antivir Res*, *100*(1), 29-37.

Maurer, G., Tarkowski, B. & Baccarini, M. (2011). Raf kinases in cancer-roles and therapeutic opportunities. *Oncogene*, *30*, 3477–3488.

Mazur, I., Wurzer, W. J., Ehrhardt, C., Pleschka, S., Puthavathana, P., Silberzahn, T., Wolff, T., Planz, O. & Ludwiq, S. (2007). Acetylsalicylic acid (ASA) blocks influenza virus propagation via its NF-kappaB-inhibiting activity. *Cell Microbiol*, *9*, 1683–1694.

Medina-Franco, J. L., Giulianotti, M. A., Welmaker, G. S. & Houghten, R. A. (2013). Shifting from the single- to the multitarget paradigm in drug discovery. *Drug Discov Today*, *18*, 495–501.

Moffat, J. F., Zerboni, L., Sommer, M. H., Heineman, T. C., Cohen, J. I., Kaneshima, H. & Arvin, A. M. (1998). The ORF47 and ORF66 putative protein kinases of varicella-zoster virus determine tropism for human T cells and skin in the SCID-hu mouse. *Proc Natl Acad Sci U S A*, *95*, 11969–11974.

Moffat, J. G., Vincent, F., Lee, J. A., Eder, J. & Prunotto, M. (2017). Opportunities and challenges in phenotypic drug discovery: an industry perspective. *Nat Rev Drug Discov*, *16*, 531–543.

Mohr, E. L., McMullan, L. K., Lo, M. K., Spengler, J. R., Bergeron, E., Albarino, C. G., Shrivastava-Ranjan, P., Chiang, C. F., Nichol, S. T., Spiropoulou, C. F. & Flint, M. (2015). Inhibitors of cellular kinases with broad-spectrum antiviral activity for hemorrhagic fever viruses. *AntivirRes*, *120*, 40-47.

Mora, A., Komander, D., van Aalten, D. M. F. & Alessi, D. R. (2004). PDK1, the master regulator of AGC kinase signal transduction. *Semin Cell Dev Biol*, *15*, 161–170.

Mueller, B. K., Mack, H. & Teusch, N. (2005). Rho kinase, a promising drug target for neurological disorders. *Nat Rev Drug Discov*, *4*, 387–398.

Narute, P. S. & Smithgall, T. E. (2012). Nef alleles from all major HIV-1 clades activate Src-family kinases and enhance HIV-1 replication in an inhibitor-sensitive manner. *PLoSOne*, *7*, e32561.

Naveja, J. J., Dueñas-González, A. & Medina-Franco, J. L. (2016). Chapter 12 - drug repurposing for epigenetic targets guided by computational methods. In: *Epi-Informatics*, ed J. L. Medina-Franco (Boston: Academic Press), 327–357.

Neveu, G., Barouch-Bentov, R., Ziv-Av, A., Gerber, D., Jacob, Y. & Einav, S. (2012). Identification and targeting of an interaction between a tyrosine motif within hepatitis C virus core protein and AP2M1 essential for viral assembly. *PLoS Pathog*, *8*, e1002845.

Neveu, G., Ziv-Av, A., Barouch-Bentov, R., Berkerman, E., Mulholland, J., & Einav, S. (2015). AP-2-associated protein kinase 1 and cyclin G-associated kinase regulate hepatitis C virus entry and are potential drug targets. *J Virol*, *89*, 4387–4404.

Noppakunmongkolchai, W., Poyomtip, T., Jittawuttipoka, T., Luplertlop, N., Sakuntabhai, A., Chimnaronk, S., Jirawatnotai, S. & Tohtong, R. (2016). Inhibition of protein kinase C promotes dengue virus replication.*Virol J*, *13*, 35.

Oprea, T. I., Nielsen, S. K., Ursu, O., Yang, J. J., Taboureau, O., Mathias, S. L.,Kouskoumvekaki, L., Sklar, L. A. & Bologa, C. G. (2011). Associating drugs, targets and clinical outcomes into an integrated network affords a new platform for computer-aided drug repurposing. *Mol Inform*, *30*, 100–111.

Ott, D. E., Coren, L. V., Johnson, D. G., Kane, B. P., Sowder, R. C., Kim, Y. D., Fisher, R. J., Zhou, X. Z., Lu, K. P. & Henderson, L. E. (2000). Actin-binding cellular proteins inside human immunodeficiency virus type 1. *Virology*, *266*, 42-51.

Pagano, M. A., Tibaldi, E., Palu, G. & Brunati, A. M. (2013). Viral proteins and Src family kinases: mechanisms of pathogenicity from a "liaison dangereuse." *World J Virol*, *2*, 71–78.

Paoletti, E. & Moss, B. (1972). Protein kinase and specific phosphate acceptor proteins associated with vaccinia virus cores. *J Virol*, *10*, 417-424.

Pauls, E., Badia, R., Torres-Torronteras, J., Ruiz, A., Permanyer, M., Riveira-Munoz, E., Clotet, B., Marti, R., Ballana, E. & Este, J. A. (2014a). Palbociclib, a selective inhibitor of cyclin-dependent kinase4/6, blocks HIV-1 reverse transcription through the control of sterile alpha motif and HD domain containing protein-1 (SAMHD1) activity. *AIDS (London, England)*, *28*, 2213–2222.

Perwitasari, O., Yan, X., O'Donnell, J., Johnson, S. & Tripp, R. A. (2015). Repurposing kinase inhibitors as antiviral agents to control influenza a virus replication. *Assay Drug Dev Technol*, *13*, 638–649.

Platz, E. A., Yegnasubramanian, S., Liu, J. O., Chong, C. R., Shim, J. S., Kenfield, S. A., Stampfer, M. J., Willett, W. C., Giovannucci, E. & Nelson, W. G. (2011). A novel two-stage, transdisciplinary study identifies digoxin as a possible drug for prostate cancer treatment. *Cancer Discov*, *1*, 68–77.

Pleschka, S., Wolff, T., Ehrhardt, C., Hobom, G., Planz, O., Rapp, U. R. & Ludwig, S. (2001). Influenza virus propagation is impaired by inhibition of the Raf/MEK/ERK signalling cascade. *Nat Cell Biol*, *3*, 301–305.

Prichard, M. N. (2009). Function of human cytomegalovirus UL97 kinase in viral infection and its inhibition by maribavir. *Rev Med Virol*, *19*, 215–229.

Prichard, M. N., Sztul, E., Daily, S. L., Perry, A. L., Frederick, S. L., Gill, R. B., Hartline, C. B., Streblow, D. N., Varnum, S. M., Smith, R. D. & Kern, E. R. (2008). Human cytomegalovirus UL97 kinase activity is required for the hyperphosphorylation of retinoblastoma protein and inhibits the formation of nuclear aggresomes. *J Virol*, *82*, 5054–5067.

Pushpakom, S., Iorio, F., Eyers, P. A., Escott, K. J., Hopper, S., Wells, A., Doig, A., Guilliams, T., Latimer, J., McNamee, C., Norris, A., Sanseau, P., Cavalla, D. & Pirmohamed, M. (2018). Drug repurposing: progress, challenges and recommendations. *Nat Rev Drug Discov*, *18*(4), 41-58.

Rechter, S., Scott, G. M., Eickhoff, J., Zielke, K., Auerochs, S., Müller, R., Stamminger, T., Rawlinson, W. D. & Marschall, M. (2009). Cyclin-dependent kinases phosphorylate the cytomegalovirus RNA export protein pUL69 and modulate its nuclear localization and activity. *J Biol Chem*, *284*, 8605–8613.

Reddy, A. S. & Zhang, S. (2013). Polypharmacology: drug discovery for the future. *Expert Rev Clin Pharmacol*, *6*, 41–47.

Reeves, P. M., Smith, S. K., Olson, V. A., Thorne, S. H., Bornmann, W., Damon, I. K. & Kalman, D. (2011). Variola and monkeypox viruses utilize conserved mechanisms of virion motility and release that depend on abl and SRC family tyrosine kinases. *J Virol*, *85*, 21–31.

Rehman, W., Arfons, L. M. & Lazarus, H. M. (2011). The rise, fall and subsequent triumph of thalidomide: lessons learned in drug development. *Ther Adv Hematol*, *2*, 291–308.

Sasaki, Y., Kakisaka, M., Chutiwitoonchai, N., Tajima, S., Hikono, H., Saito, T. &Aida, Y. (2014). Identification of a novel multiple kinase inhibitor with potent antiviral activity against influenza virus by reducing viral polymerase activity. *Biochem Biophys Res Commun*, *450*(1), 49-54.

Sato, S., Sanjo, H., Takeda, K., Ninomiya-Tsuji, J., Yamamoto, M., Kawai, T., Matsumoto, K., Takeuchi, O. & Akira, S. (2005). Essential function for the kinase TAK1 in innate and adaptive immune responses. *Nat Immunol*, *6*, 1087–1095.

Schang, L. M. (2003). The cell cycle, cyclin-dependent kinases, and viral infections: new horizons and unexpected connections. *Prog Cell Cycle Res*, *5*, 103–124.

Schang, L. M., St Vincent, M. R., & Lacasse, J. J. (2006). Five years of progress on cyclindependent kinases and other cellular proteins as potential targets for antiviral drugs. *Antivir Chem Chemother*, *17*, 293-320.

Schang, L. M., St. Vincent, M. R. & Lacasse, J. J. (2006). Five years of progress on cyclin-dependent kinases and other cellular proteins as potential targets for antiviral drugs. *Antiviral Chem Chemother*, *17*, 293–320.

Scheuch, G., Canisius, S., Nocker, K., Hofmann, T., Naumann, R., Pleschka, S., Ludwiq, S., Welte, T. & Plan, O. (2018). Targeting intracellular signaling as an antiviral strategy: aerosolized LASAG for the treatment of influenza in hospitalized patients. *Emerg Microbes Infect*, *7*(1), 21.

Schleiss, M., Eickhoff, J., Auerochs, S., Leis, M., Abele, S., Rechter, S., Choi, Y., Anderson, J., Scott, G., Rawlinson, W., Michel, D., Ensminger, S., Klebl, B., Stamminger, T. & Marschall, M. (2008). Protein kinase inhibitors of the quinazoline class exert anti-cytomegaloviral activity *in vitro* and *in vivo*. *Antivir Res*, *79*, 49–61.

Schor, S. & Einav, S. (2018). Repurposing of kinase inhibitors as broad-spectrum antiviral drugs. *DNA Cell Biology*, *37*(2), 63-69

Schräder, T., Dudek, S. E., Schreiber, A., Ehrhardt, C., Planz, O. & Ludwig, S. (2018). The clinically approved MEK inhibitor Trametinib efficiently blocks influenza A virus propagation and cytokine expression. *Antivir Res*, *157*, 80–92.

Sen, A. (1981). Purified low-molecular-weight protein kinase from murine sarcoma virus particles catalyzes tyrosine phosphorylation endogenously but phosphorylates cellular proteins at serine. *J Virol*, *39*, 612-624.

Sen, A. & Todaro, G. J. (1979). A murine sarcoma virus-associated protein kinase: interaction with actin and microtubular protein. *Cell*, *17*, 347-356.

Sen, A., Todaro, G. J., Blair, D. G. & Robey, W. G. (1979). Thermolabile protein kinase molecules in a temperature-sensitive murine sarcoma virus pseudotype. *Proc Natl Acad Sci USA*, *76*, 3617-3621.

Shives, K. D., Beatman, E. L., Chamanian, M., O'Brien, C., Hobson-Peters, J. & Beckham, J. D. (2014). West nile virusinduced activation of mammalian target of rapamycin complex 1 supports viral growth and viral protein expression. *J Virol*, *88*, 9458–9471.

Sidders, B., Karlsson, A., Kitching, L., Torella, R., Karila, P., Phelan, A. (2018). Network based drug discovery: coupling network pharmacology with phenotypic screening for neuronal excitability. *J Mol Biol*, *430*, 3005–3015.

Slater, M. J., Cockerill, S., Baxter, R., Bonser, R. W., Gohil, K., Gowrie, C., Robinson, J. E., Littler, E., Parry, N., Randall, R. & Snowden, W. (1999). Indolocarbazoles: potent, selective inhibitors of human cytomegalovirus replication. *Bioorg Med Chem*, *7*(6), 1067-1074.

Steingruber, M., Kraut, A., Socher, E., Sticht, H., Reichel, A., Stamminger, T., Amin, B., Couté, Y., Hutterer, C. & Marschall, M. (2016). Proteomic Interaction Patterns between Human Cyclins, the Cyclin-Dependent Kinase Ortholog pUL97 and Additional Cytomegalovirus Proteins. *Viruses*, *8*(8), pii: E219.

Steingruber, M., Socher, E., Hutterer, C., Webel, R., Bergbrede, T., Lenac, T., Sticht, H. & Marschall. M. (2015). The Interaction between Cyclin B1 and Cytomegalovirus Protein Kinase pUL97 is Determined by an Active Kinase Domain. *Viruses*, *7*(8), 4582-4601.

Stephenson, S. T., Bostik, P., Song, B., Rajan, D., Bhimani, S., Rehulka, P., Mayne, A. E. &Ansari, A. A. (2011). Distinct host cell proteins incorporated by SIV replicating in CD4+ T cells from natural disease resistant versus non-natural disease susceptible hosts. *Retrovirology*, *7*, 107.

Stokoe, D., Stephens, L., Copeland, T., Gaffney, P., Reese, C., Painter, G., Holmes, A., McCormick, F. & Hawkins, P. (1997). Dual role of phosphatidylinositol-3,4,5-trisphosphate in the activation of protein kinase B. *Science*, 567–570.

Strand, M. & August, J. T. (1971). Protein kinase and phosphate acceptor proteins inRauscher murine leukemia virus. *Nature New Biol*, *233*, 137-140.

Strittmatter, S. M. (2014). Overcoming drug development bottlenecks with repurposing: old drugs learn new tricks. *Nat Med*, *20*, 590–591.

Sullivan, V., Talarico, C. L., Stanat, S. C., Davis, M., Coen, D. M. & Biron, K. K. (1992). A protein kinase homologue controls phosphorylation of ganciclovir in human cytomegalovirus-infected cells. *Nature*, *358*, 162–164.

Supekova, L., Supek, F., Lee, J., Chen, S., Gray, N., Pezacki, J. P., Schlapbach, A. & Schultz, P. G. (2008). Identification of human kinases involved in hepatitis C virus replication by small interference RNA library screening. *J Biol Chem*, *283*, 29–36.

Swimm, A. I., Bornmann, W., Jiang, M., Imperiale, M. J., Lukacher, A. E. & Kalman, D. (2010). Abl family tyrosine kinases regulate sialylated ganglioside receptors for polyomavirus. *J Virol*, *84*, 4243–4251.

Swinney, D. C. & Anthony, J. (2011). How were new medicines discovered? *Nat Rev Drug Discov*, *10*, 507–519.

Tabit, C. E., Shenouda, S. M., Holbrook, M., Fetterman, J. L., Kiani, S., Frame, A. A., Kluge, M. A., Held, A., Dohadwala, M. M. & Gokce, N. (2013). Protein kinase C-β contributes to impaired endothelial insulin signaling in humans with diabetes mellitus. *Circulation*, *127*, 86–95.

Tan, K. B. (1975). Comparative study of the protein kinase associated with animal viruses. *Virology*, *64*, 566–570.

Tardif, M. R., Gilbert, C., Thibault, S., Fortin, J. F. & Tremblay, M. J. (2009). LFA-1 antagonists as agents limiting human immunodeficiency virus type 1 infection and transmission and potentiating the effect of the fusion inhibitor T-20. Antimicrob Agents *Chemother*, *53*, 4656-4666.

Tari, L. B. & Patel, J. H. (2014). Systematic drug repurposing through text mining. *Methods Mol Biol*, *1159*, 253–267.

Thomas, A. B. (2006). The role of serendipity in drug discovery. *Dialogues Clin Neurosci*, *8*(3), 335-344.

Tidona, C. A. & Darai, G. (1997). The complete DNA sequence of lymphocystis disease virus. *Virology*, *230*, 207–216.

Trofe, J., Pote, L., Wade, E., Blumberg, E. & Bloom, R. D. (2008). Maribavir: a novel antiviral agent with activity against cytomegalovirus. *Ann Pharmacother*, *42*, 1447–1457.

Tsiapalis, C. M. (1977). Protein kinase and phosphoproteins of avian myeloblastosis virus. *J Virol*, *21*, 843-848.

Ueki, I. F., Min-Oo, G., Kalinowski, A., Ballon-Landa, E., Lanier, L. L., Nadel, J. A. & Koff, J. L. (2013). Respiratory virus–induced EGFR activation suppresses IRF1-dependent interferon l and antiviral defense in airway epithelium. *J Exp Med*, *210*, 1929–1936.

van der Graaf, P. H. & Benson, N. (2011). Systems pharmacology: bridging systems biology and pharmacokinetics-pharmacodynamics (PKPD) in drug discovery and development. *Pharm Res*, *28*, 1460–1464.

Vigne, S., Duraffour, S., Andrei, G., Snoeck, R., Garin, D. & Crance, J.M. (2009). Inhibition of vaccinia virus replication by two small interfering RNAs targeting B1R and G7L genes and their synergistic combination with cidofovir. *Antimicrob Agents Chemother*, *53*, 2579–2588.

Waldman, W. J., Knight, D. A., Blinder, L., Shen, J., Lurain, N. S., Miller, D. M., Sedmak, D. D., Williams, J. W. & Chong, A. S. (1999). Inhibition of cytomegalovirus *in vitro* and *in vivo* by the experimental immunosuppressive agent leflunomide. *Intervirology*, *42*, 412–418.

Wang, X., Huang, D. Y., Huong, S. M. & Huang, E.S. (2005). Integrin avb3 is a Coreceptor for Human Cytomegalovirus Infection. *Nat Med*, *11*, 515–521.

Wang, X., Huong, S.M., Chiu, M. L., Raab-Traub, N. & Huang, E.S. (2003). Epidermal growth factor receptor is a cellular receptor for human cytomegalovirus. *Nature*, *424*, 456–461.

Watanabe, T., Kawakami, E., Shoemaker, J. E., Lopes, T. J. S., Matsuoka, Y., Tomita, Y., Kozuka-Hata, H., Gorai, T., Kuwahara, T., Takeda, E., Nagata, A., Takano, R., Kiso, M., Yamashita, M., Sakai-Tagawa, Y., Katsura, H., Nonaka, N., Fujii, H., Fujii, K., Sugita, Y., Noda, T., Goto, H., Fukuyama, S., Watanabe, S., Neumann, G., Oyama, M., Kitano, H. & Kawaoka, Y. (2014). Influenza virus-host interactome screen as a platform for antiviral drug development. *Cell Host Microbe*, *16*, 795–805.

Weller, M. L., Amornphimoltham, P., Schmidt, M., Wilson, P. A., Gutkind, J. S. & Chiorini, J. A. (2010). Epidermal growth factor receptor is a co-receptor for adeno-associated virus serotype 6. *Nat Med*, *16*, 662–664.

Williams, J. W., Javaid, B., Kadambi, P. V., Gillen, D., Harland, R., Thistlewaite, J. R., Garfinkel, M., Foster, P., Atwood, W., Millis, J. M., Meehan, S. M. & Josephson M. A. (2005). Leflunomide for polyomavirus type BK nephropathy. *N Engl J Med*, *352*, 1157–1158.

Wurzer, W. J., Ehrhardt, C., Pleschka, S., Berberich-Siebelt, F., Wolff, T., Walczak, H., Planz, O. & Ludwiq, S. (2004). NF-kappaB-dependent induction of tumor necrosis factorrelated apoptosis-inducing ligand (TRAIL) and Fas/FasL is crucial for efficient influenza virus propagation. *J Biol Chem*, *279*, 30931–30937.

Xu, M., Lee, E. M., Wen, Z., Cheng, Y., Huang, W.-K., Qian, X., Tcw, J., Kouznetsova, J., Ogden, S. C., Hammack, C., Jacob, F., Nguyen, H. N., Itkin, M., Hanna, C., Shinn, P., Allen, C., Michael, S. G., Simeonov, A., Huang, W., Christian, K. M., Goate, A., Brennand, K. J., Huang, R., Xia, M., Ming, G. L., Zheng, W., Song, H. & Tang, H. (2016). Identification of small-molecule inhibitors of Zika virus infection and induced neural cell death via a drug repurposing screen. *Nat Med*, *22*, 1101–1107.

Yamamoto, M., Onogi, H., Kii, I., Yoshida, S., Iida, K., Sakai, H., Abe, M., Tsubota, T., Ito, N., Hosoya, T. & Hagiwara, M. (2014). CDK9 inhibitor FIT-039 prevents replication of multiple DNA viruses. *J Clin Invest*, *124*, 3479–3488.

Yang, X. & Gabuzda, D. (1999). Regulation of human immunodeficiency virus type 1 infectivity by the ERK mitogen-activated protein kinase signaling pathway. *J Virol*, *73*, 3460–3466.

Yin, M. J., Yamamoto, Y. & Gaynor, R. B. (1998). The anti-inflammatory agents aspirin and salicylate inhibit the activity of I(kappa)B kinase-beta. *Nature*, *396*, 77–80.

Zampieri, C. A., Fortin, J.-F., Nolan, G. P. & Nabel, G. J. (2007). The ERK Mitogen-activated protein kinase pathway contributes to Ebola virus glycoprotein-induced cytotoxicity. *J Virol*, *81*, 1230–1240.

Zheng, K., Kitazato, K. & Wang, Y. (2014). Viruses exploit the function of epidermal growth factor receptor. *Rev Med Virol*, *24*, 274–286.

Zhou, W., Wang, Y., Lu, A. & Zhang, G. (2016). Systems pharmacology in small molecular drug discovery. *Int J Mol Sci*, *17*, 246.

Zhu, J., Huang, J. W., Tseng, P.H., Yang, Y.T., Fowble, J., Shiau, C. W., Shaw, Y. J., Kulp, S. K. & Chen, C.S. (2004). From the cyclooxygenase-2 Inhibitor celecoxib to a novel class of 3-phosphoinositide-dependent protein kinase-1 inhibitors. *Cancer Res*, *64*, 4309–4318.

Zimmermann, A., Wilts, H., Lenhardt, M., Hahn, M. & Mertens, T. (2000). Indolocarbazoles exhibit strong antiviral activity against human cytomegalovirus and are potent inhibitors of the pUL97 protein kinase. *Antivir Res*, *48*(1), 49-60.

Zona, L., Lupberger, J., Sidahmed-Adrar, N., Thumann, C., Harris, Helen J., Barnes, A., Florentin, J., Tawar, Rajiv G., Xiao, F., Turek, M.,Durand, S. C., Duong, F. H., Heim, M. H., Cosset, F. L., Hirsch, I., Samuel, D., Brino, L., Zeisel, M. B., Le Naour, F., McKeating, J. A. & Baumert, T. F. (2013). HRas signal transduction promotes hepatitis C virus cell entry by triggering assembly of the host tetraspanin receptor complex. *Cell Host Microbe*,*13*, 302–313.

In: A Closer Look at Kinase Inhibition ISBN: 978-1-53616-973-7
Editor: Alex Tompson © 2020 Nova Science Publishers, Inc.

Chapter 3

IN SEARCH OF NOVEL TAU KINASE INHIBITORS AND THEIR ROLE IN ALZHEIMER'S THERAPY

Arvind Kumar Jain[1], Saloni Malla[2], C. Karthikeyan[3], Amit K. Tiwari[2], Piyush Trivedi[4] and Anita Dutt Konar[1,5,*]

[1]School of Pharmaceutical Sciences, Rajiv Gandhi Technological University, Bhopal, Airport Bypass Road, Gandhinagar, Bhopal, India
[2]Department of Pharmacology and Experimental Therapeutics, College of Pharmacy and Pharmaceutical Sciences, University of Toledo, OH, US
[3]Department of Pharmacy, Indira Gandhi National Tribal University, Amarkantak (MP), India
[4]Centre of Innovation and Translational Research, Poona College of Pharmacy, Bharati Vidyapeeth Deemed University, Pune, India
[5]Departmentof Applied Chemistry Rajiv Gandhi Technological University, Bhopal

[*] Corresponding Author's Email: anitaduttkonar@rgtu.net.

ABSTRACT

Ubiquitously expressed, kinases are key enzymes that regulate the cellular metabolism by catalyzing the phosphorylation of serine/threonine/tyrosine residues by transferring a γ-phosphate moiety from ATP. The dysregulation or an imbalance in kinase signaling pathways can produce pathological changes that result in disease. For example, Alzheimer's (AD), the most common neurodegenerative disease, is usually caused by hyper-phosphorylation of Tau kinases. Tau is a microtubule-associated protein that is significantly expressed in healthy neurons and plays a significant role in the neuronal cytoskeleton stabilization. Structurally, the Tau protein is divided into four domains: an acidic region in the N-terminal part, a proline-rich region, a microtubule-binding domain (MBD) containing four repeat units and a C-terminal region respectively. Interestingly, each repeat domain in the MBD contains a conserved consensus motif KXGS, which is readily phosphorylated, resulting in the destabilization of the neuronal cytoskeleton. Consequently, phosphorylation disrupts Tau-dependent cellular functions, thereby promoting Tau dissociation from microtubules, forming neurofibrillary tangles (NFT), a pathological hallmark of AD. Recent studies have shed light on the pathological phosphorylation sites responsible for Tau aggregation. This information has led to research involving the development of inhibitors targeting these kinases. In AD, Tau phosphorylation precedes Tau aggregation and the subsequent formation of NFT. Several families of kinases are known to be involved and are well documented and reviewed elsewhere. This chapter focuses on the key kinases, namely CLK1, DYRK1A, CDK5, GSK3β and CK1δ as data indicate that targeting these proteins may have therapeutic promise against AD. Moreover o*ptimization techniques,* including chemical modifications of small heterocyclic nucleus, that are *currently being used for* the development of highly potent, safe and selective inhibitors has been discussed here.

BRIEF OUTLINE ABOUT ALZHEIMER'S DISEASE

Alzheimer's disease the most prevalent neurodegenerative disease, has gained significant impetus in the last few decades amongst the elderly population. Evidences reveal that this disorder mainly originates owing to the accumulation of intracellular deposites of microtubule associated tau proteins into neurofibrillar tangles (NFTs) at specific regions of the brains

[1-3]. Indeed the control of microtubule stabilization depends upon a number of post-translational modification pathways at specific amino acid residues namely phosphorylation. Slight increase in number of the phosphorylation units from the threshold, leads to abrupt neuronal cytoskeleton destabilization, affecting normal cell metabolism [1-3].

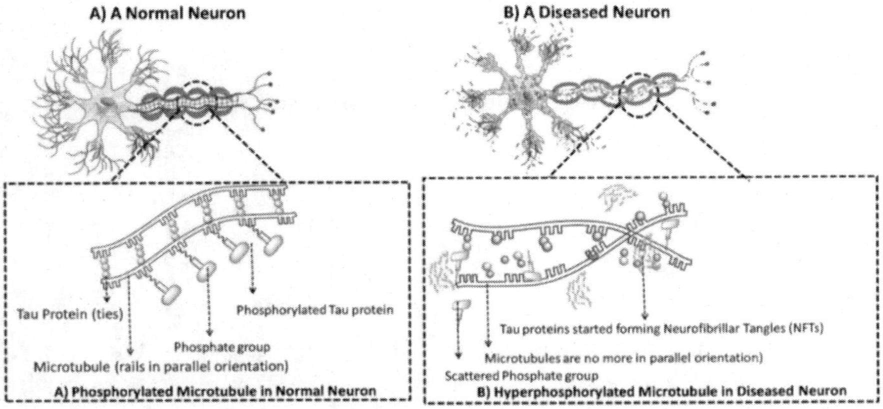

Figure 1. The probable internal structure of a healthy and diseased neuron.

In order to understand the regulation of normal cell metabolism, a closer inspection of the structural aspect of the neurons needs to be addressed. Basically, the hall marks of neurons are the axons that function as a cable, electrically wiring the brain and connecting the brain to the body [1-3]. This completed circuit allows a person to react to his/her environment in response to a stimulus or stimuli (Figure 1). The axon primarily consists of microtubules (MTs) that are tightly held together by Tau proteins so that they remain parallel to each other [1-3]. This permits smooth transport of nutrients and other essential biomolecules from the cell body of the neuron (soma) to the axon tip. Overall, the axons are interconnected to the nearby axons, forming a closed circuit and allowing normal neuronal cell function to occur.To understand the mechanism in a simplified manner, the microtubules have been compared to a pair of rails on a railroad. For normal cell function, these rails must be held together in a parallel position by the ties, that are analogous to tau proteins (Figure 1A). When these ties (Tau proteins) cannot keep the rails (microtubules) in a parallel orientation, the former starts aggregating as insoluble deposits, known as neurofibrillary tangles (NFTs) (Figure 1B). Consequently the transport of biomolecules from the cell body to the axon tip is impeded leading to shrinkage of the cellular organalles. Thus the aforementioned pathological process results in the onset of Alzheimer's disease (Figure 1) [4].

Why Do The Ties (Tau Proteins) Fail to Keep The Rails (MTs) in Parallel Orientation?

The Tau proteins, which hold the microtubules in the appropriate orientation, are regulated by several post translational modifications, with the most common being phosphorylation. However, this process is normally regulated by a delicate balance between the activities of two enzymes, namely kinases and phosphatases [5]. Kinases catalyze the transfer of γ-phosphate from ATP to its protein substrates, whereas phosphatases removes phosphates from the phosphorylated proteins. The maintenance of a steady state between two opposing mechanisms, modulates the conformational and functional aspects of cellular proteins. Normally, a native protein in the human brain on an average has two phosphate groups, whereas tau in diseased brain can have up to 8 phosphorylated residues (Figure 2) [5]. This abnormal phosphorylation (i.e., hyperphosphorylation) disrupts the kinase - phosphatase balance, initiating the development of paired helical filamants, followed by straight fibers and ultimately NFTs in specific regions of the brain (Figure 2) [1, 4, 6, 7]. In other words the tau proteins cannot maintain the parallel orientation of the microtubules which decreases or even blocks the transport of food and nutrients to the axon tip. Consequently, the circuit connection becomes incomplete, causing a shrinkage of the nerve cell.

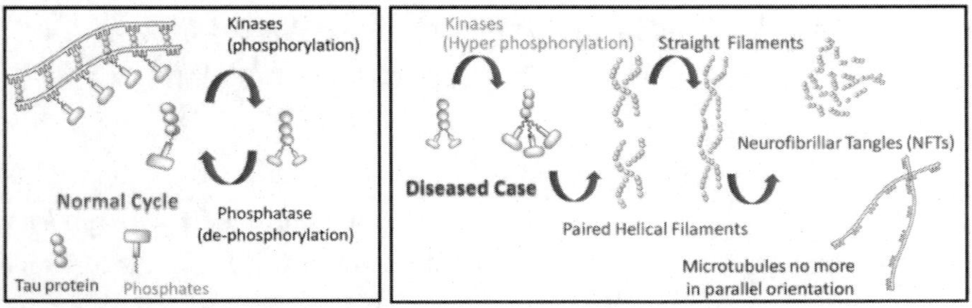

Figure 2. The phosphorylation-dephosphorylation pathway in normal cycle and diseased case.

What Are Protein Kinases and What Do They Do?

Protein kinases are a group of enzymes that catalyze the transfer of a phosphate group to proteins, which significantly alters bioactivity, as well their interaction with certain biomolecules. In this process, ATP or GTP is the phosphate donor, while the hydroxy substituent of serine/threonine and tyrosine are the acceptors (Scheme 1) [8]. As the member of cell signal chemical modification, they are particularly prominent in signal transduction and is known to regulate majority of the cell processes, such as metabolism, transcription, cell cycle progression, cell movement, apoptosis, and differentiation [8]. Moreover numerous studies indicate that they are involved in etiology of cancer, diabetes, autoimmune diseases and several others [8]. Therefore, it has been postulated that the modulation of the activity of the protein kinases is a potential strategy for the treatment of specific diseases [8].

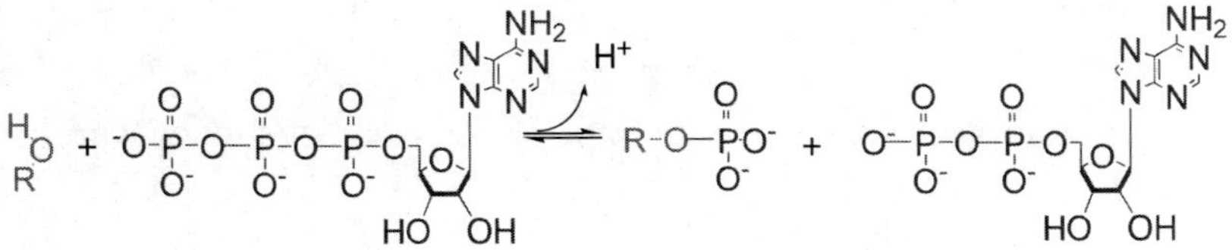

Scheme 1. γ-Phosphoryl transfer reaction from ATP.

Are Any Single Serine/Threonine Residue or All Serine/Threonine Residues in Receptors Phosphorylable?

To answer this question basic conception of "Consensus Sequence" is necessary [9-10]. From literature we learnt that before kinase begins its activity it recognizes and binds to the amino acids towards both the N and C- termini, at the active site. These terminal sequences to which the kinases binds before becoming operational are the consensus sequences. This event therefore positions the phosphorylable residues especially Ser/Thr/Tyr in close proximity to the ATPs/GTPs, permitting smooth phosphate transfer. If the residues donot fall in the proximity of the consensus sequences it do not undergo phosphorylation. The basic characterestic of these sequences depends on the polarity and nature of the amino acids [9-10]. Thus the amino acid to be phosphorylated and its corresponding consensus sequences are two important parameters that governs the categorization of the protein kinase family as described in the subsequent section below.

How Are Kinases Classified?

The molecular phylogeny of the protein kinases upon which Kinome is based was first described by Hanks and Hunter in 1991 [10-11]. In their classification scheme the eukaryotic protein kinase superfamily comprising of 518 members were broadly split into two groups: 'Conventional' (**ePK**) (478 members) and 'A typical' protein kinases (**aPKs**) (40 members) (Figure 3). **ePKs** are the largest group, and have been sub-classified into 9 members by examining sequence similarity between the catalytic and accessory domains in addition to their modes of regulation. The **aPKs** are a small set of protein kinases that do not share clear sequence similarity with ePKs [12].

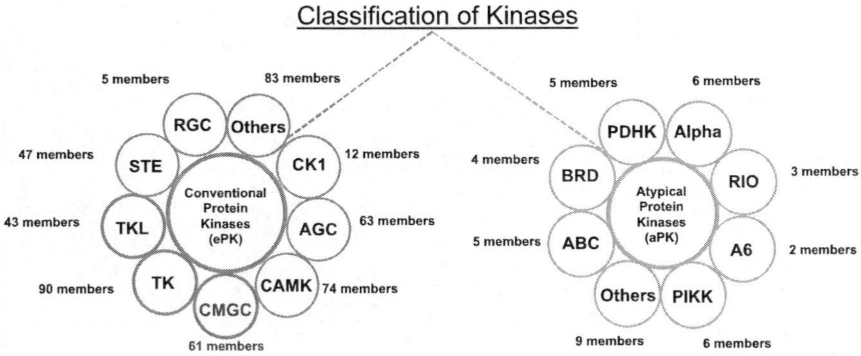

Figure 3. The kinase family with all its members.

The aberrant hyperphosphorylation of tau has been postulated to be involved in the pathogenesis of Alzheimer's disease, leading to the formation of tau aggregates in specific areas of the brain [1-4]. Therefore, designing molecules that reduce tau aggregation could represent a valid therapeutic strategy for Alzheimer's, as well as other related neurodegenerative diseases [1-4]. In this book chapter our focus lies to report the design of small molecule inhibitors of few members of the CMGC family namely DYRK1A, CLK1, CDK5, GSK3β and CK1δ, which falls in the class of ePKs. This is because these members have been directly involved in the hyperphosphorylation of the microtubule associated tau proteins [7].

Why CMGC?

The CMGC family, comprising of 61 members, are an important class of kinases whose activities govern cell signaling pathways [11-12]. Their name originates from the initials of the subfamily members and their respective functions are as described: CDK (cyclin-dependent kinase: responsible for regulating the cell cycle progression); MAPK (mitogen-activated protein kinase: for driving cellular processes); GSK (glycogen synthase kinase: for controlling glycogen metabolism) and CLK (CDC like kinase, to control RNA splices) (Figure 4) [11-12].

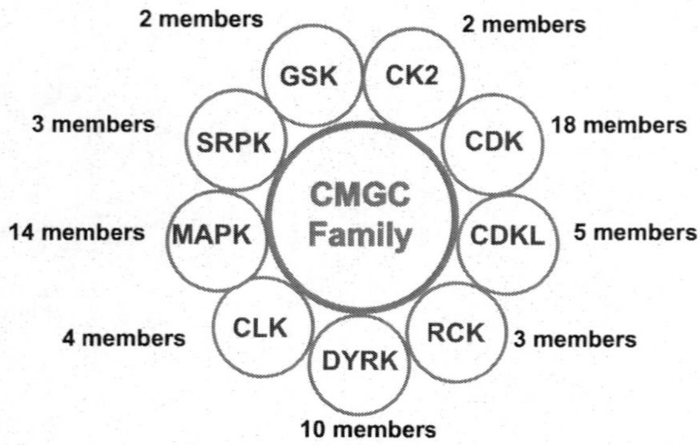

Figure 4. Detailed classification of the CMGC Family Members.

Why These Selected Members *DYRK1A, CLK1, CDK5, GSK3β* and *CK1δ?*

Literature documentation reveals that all these members DYRK1A, CLK1, CDK5, GSK3β and CK1δ are directly involved in hyperphosphorylation of microtubule binding protein Tau [7]. Apart from these there are some specific reasons lying behind choosing these targets. They are:

a) DYRK1A has received much attention due to its implication in the development of sporadic Alzheimer's disease and Down syndrome [13-14].
b) CLKs share a strong structural homology and physiological functions with the DYRK family [15-16]. Moreover they are specifically involved in patho physiology of Alzheimer's by phosphorylating selectively the serine residue in Ser/ Arg rich proteins.
c) The activities of GSK3β and CDK5 are interdependant on each other [17]. In young age, p-25 subunit of CDK5 activates the target CDK5, by inhibiting phosphorylation at Ser 9 residue of GSK3β, allowing normal cell metabolism. But in older age, this p-

25 unit of CDK5 itself becomes diseased (hyperphosphorylated). Henceforth the latter fails to display this inhibition process of phosphorylation at GSK3β, thereby leading to the onset of Alzheimer's disease. This data of interdependency has been best described by Professor Giese as "Both the CDK5 and GSK3β bear dynamic cross talk with each other" [17].

d) Considering the case of CK1δ, it has been found to be an important determinant in the development and progression of Alzheimer's [18].

Thus to design potent and selective aforementioned inhibitors, it is critical to understand the important structural aspects of the target enzymes. Therefore, we delved deeper into the crystal structure of the catalytic domain of: A) General Protein Kinases (Figure 5); and B) The respective targets with their co-crystallized ligands (Figure 6 – 10).

Structural Aspect of Catalytic Domain of General Protein Kinases

As reflected from the crystal structure, the catalytic domain of kinases comprises of two major lobes: a smaller N-terminal ''lobe'' embedded with five *β*-strands and the larger C-terminal lobe with a critical α-helix unit (Figure 5) [8, 19]. These two subdomains are interconnected by a peptide strand, forming a cleft, preferably known as the hinge region.

This cleft consists of two different pockets. a) The front pocket or the ATP binding region constituting the active site. Here several interactions occur which has been described sequentially. Firstly the adenine of ATP forms H-bond with the amide backbone of the hinge region (Figure 5B) [8, 19]. Secondly, the activation loop is buried deep in the dynamic P-segment, the activation state of which depends upon the approach of the incoming ligands (Figure 5A) [8, 19]. Thirdly, here lies the conserved DFG sequence, where the aspartate residue binds directly to the magnesium ion cofactor, concomittantly orienting the phosphate of ATP for transfer (Figure

5B) [8, 19]. Fourthly the consensus sequence exists here which is directly involved in ionic interactions with the phosphate groups of ATP (Figure 5B) [8, 19]. Indeed incorporation of the phosphate group creates a network of strong hydrogen bonds with the residues of the active site and the phosphoacceptor binding region, in addition to the former ionic interactions. Under such circumstance the conserved lysine and gate keeper residue orients the ATP for catalysis and permits the access of the hydrophobic region of the catalytic domain to the incoming ligands [8, 19]. This correspondence is responsible for introducing selectivity in the inhibitors; b). The other pocket of the cleft is the back pocket that supports the regulatory functions (Figure 5) [8, 19]. In inactive conformations, the C-terminus is rotated away from the ATP binding site. But during the activation state, the C-terminal lobe and the gatekeeper residue face towards each other [20].

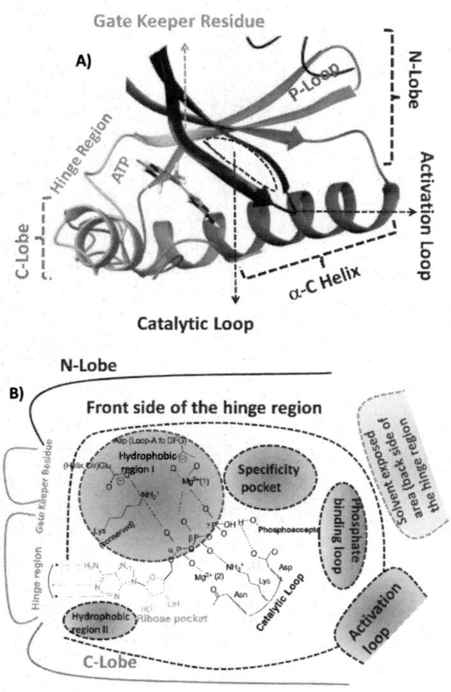

Figure 5. A) 3D and B) 2D Structures of the Catalytic domain of Kinases responsible for introducing selectivity and potency in the inhibitors.

In the foregoing section, we have discussed the primary interactions at the active site of the mentioned receptors with their co-crystallized ligand obtained from the crystal structure from the Protein Data Bank (PDB data bank).

The Design of Inhibitors for the Receptor DYRK1A

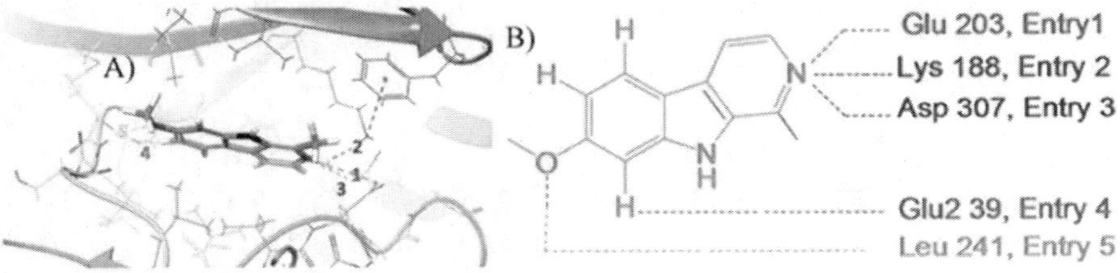

Figure 6. A; Molecular modelling 3D image of the reference compound Harmine, co-crystallised within the ATP binding site of the receptor DYRK1A. B: Structure of the reference compound demonstrating various interactions with the residues of the active site. The water molecules have been omitted for clarity. The reference ligand is shown in green. The various interactions with different residues (shown in various colours) observed at the active site are shown in dotted lines.

To investigate the possible interactions at the ATP binding site of the DYRK1A sequence, herein we report the co-crystallised structure of the reference ligand, harmine, modeled in the active site using the PDB Grid: 3ANR (Figure 6) [21]. The major interactions that were observed are as follows: 1) A salt bridge formation between the pyridinyl nitrogen of harmine with hydroxy group of acidic side chain of Glutamic acid (Glu 203) (Table 1, Entry 1) (Figure 6B); 2) A H-bond with side chain NH of Lys 188 with pyridinyl nitrogen of harmine (Table 1, Entry 2) (Figure 6B); 3) A second salt bridge with the same pyridinyl nitrogen, this time with hydroxy group of acidic side chain of Aspartic acid (Asp 307) (Table 1, Entry 3) (Figure 6B); 4) A second H-bond with backbone carbonyl of Gate keeper residue glutamic acid (Glu 239) with CH of benzene ring of indole at the same side of indole NH (Table 1, Entry 4) (Figure 6B); 5) A third H-bond with methoxy oxygen of indole with Leu 241 NH. (Table 1, Entry 5) (Figure 6B).

Table 1. Various interactions of the reference ligand Harmine within the active site of the receptor DYRK1A

Entry	Region of interaction	Interactions involved in the active site	Residues involved	
			Donor (Source)	Acceptor
1.	Phosphate Binding Region (PBR)	Salt bridge	Pyridinyl H of Harmine	Glu 203 side chain OH
2.	Solvent exposed area	H-bond	Lys 188 side chain NH	Pyridinyl N of Harmine
3.	Phosphate Binding Region (PBR)	Salt bridge	Pyridinyl H of Harmine	Asp 307 side chain OH
4.	Gate Keeper Residue	H-bond	C-H of benzene ring of indole at the same side of indole NH	Glu 239 backbone CO
5.	Hinge region	H-bond	Leu 241 backbone NH	Harmine OMe

In the subsequent section, we report a series of DYRK1A inhibitors that have shown considerable potency and selectivity towards this target from the last decade to till date (micro molar to nano molar).

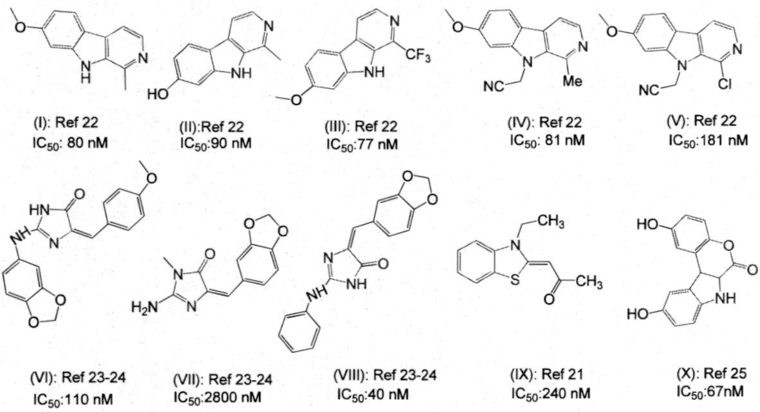

Scheme 2. Structures of some of the selected inhibitors of the target DYRK1A with their IC_{50} values. The pharmacophore responsible for inducing the activity is coloured in blue.

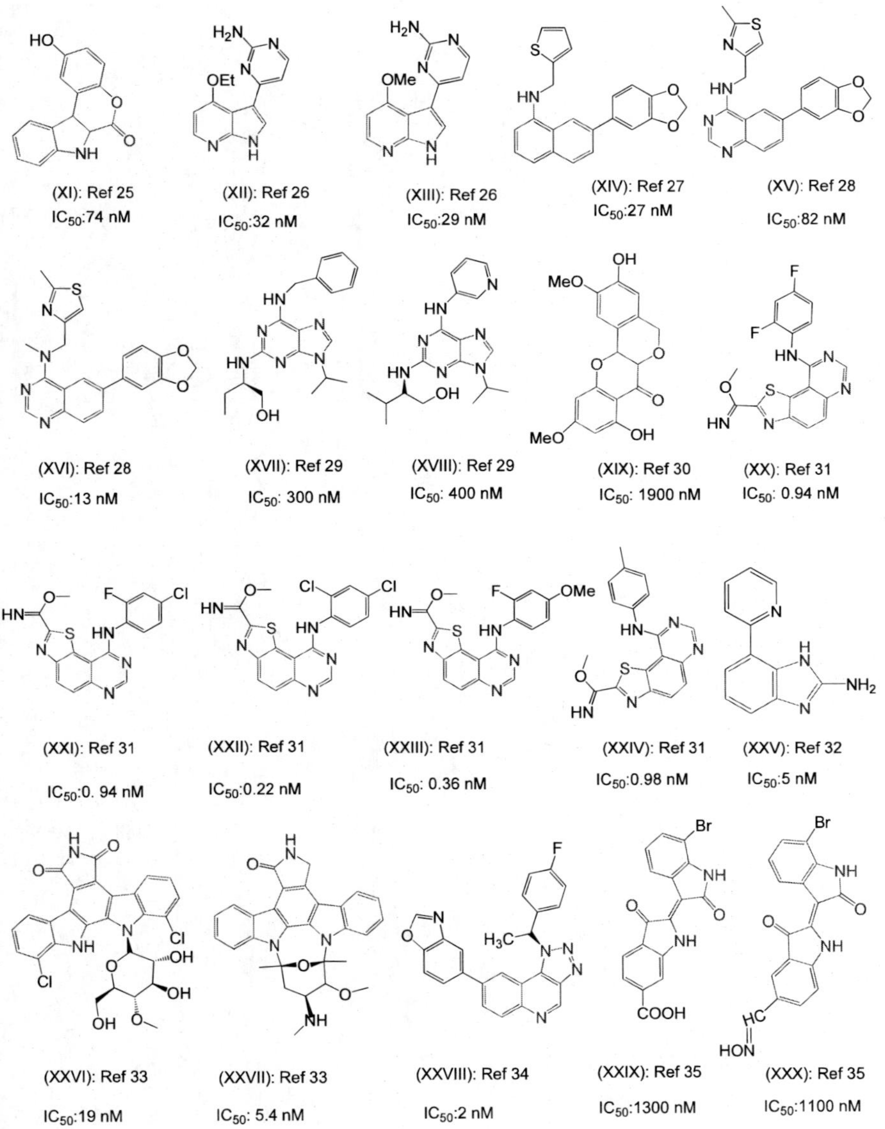

Scheme 3. Structures of some more selected inhibitors of the target DYRK1A with their IC_{50} values. The pharmacophore responsible for inducing the activity is coloured in blue.

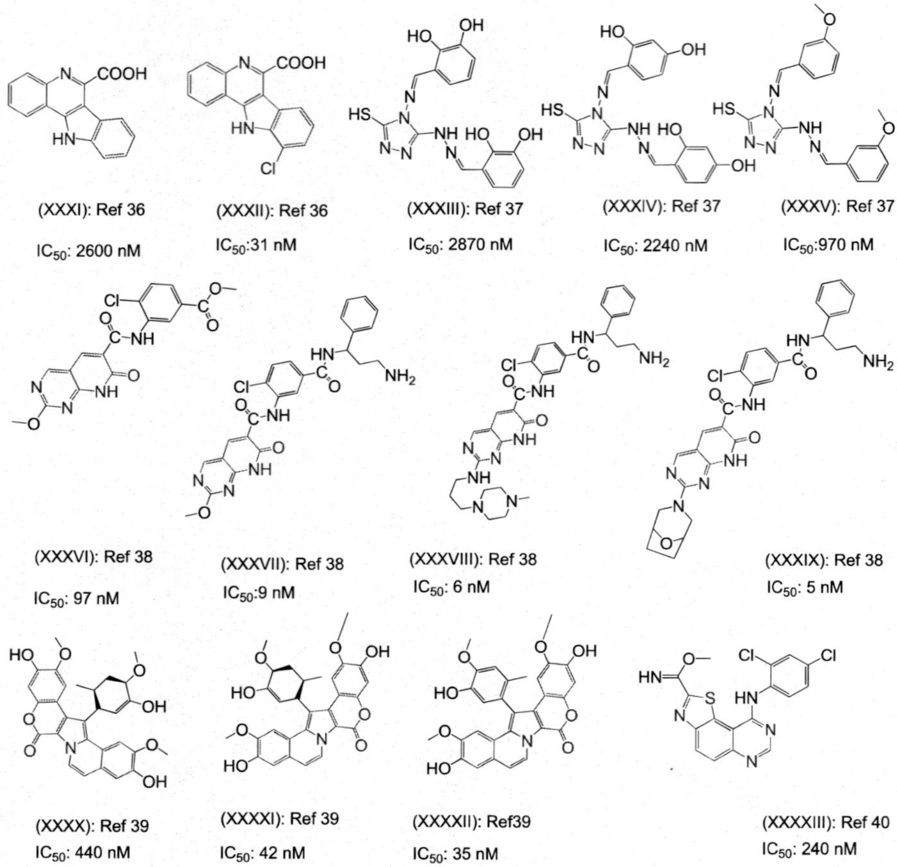

Scheme 4. Few more selected inhibitors of the target DYRK1A with their IC_{50} values. The pharmacophore responsible for inducing the activity is coloured in blue.

Design of Inhibitors for the Receptor CLK1

To determine the plausible interactions at the ATP binding site of the receptor CLK1, we report the co-crystallised structure of the reference ligand, debromo hymenialdisine, docked in the active site,using the PDB Grid: 1Z57 (Figure 7) [41]. The correspondences that were observed are as follows: 1) A H-bonding interaction with NH of Leu 244 with Azepine carbonyl of debromo hymenialdisine (Table 2, Entry 1) (Figure 7); 2) A second H-bond between azepine NH of debromo hymenialdisine and CO of

Glutamic acid (Glu 242) (Table 2, Entry 2) (Figure 7); 3) A Third H-bond with the N adjacent to carbonyl of imidazole ring of debromo hymenialdisine with Lys 191 side chain NH (Table 2, Entry 3) (Figure 7); 4) A Fourth H- bond with NH_2 of imidazole with side chain OH of Asp325 (Table 2, Entry 4) (Figure 7).

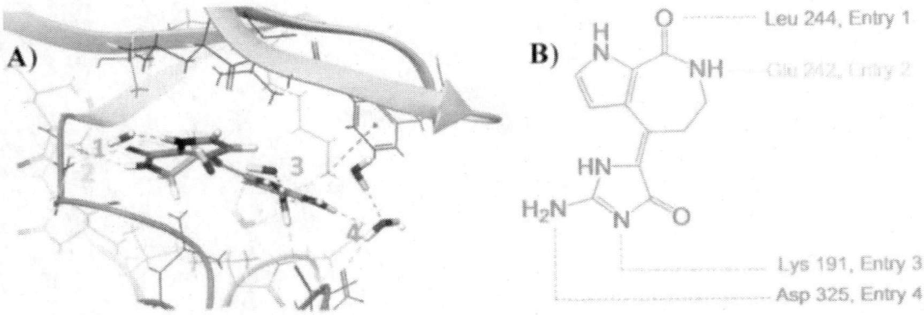

Figure 7. A; Molecular modelling 3D image of the reference compound debromo-hymenaldisine, within the ATP binding site of the receptor CLK1. B: Structure of the reference compound demonstrating various interactions within the residues of the active site. The water molecules have been omitted for clarity. The reference ligand is shown in green. The various interactions with different residues (shown in various colours) observed at the active site are shown in dotted lines.

Table 2. Various interactions of the reference ligand Hymenaldesine (PDB ID: 1Z57) with CLK1

Entry	Region of interaction	Interactions involved in the active site	Residues involved	
			Donor (Source)	Acceptor
1.	Hinge region	H-bond	Leu 244 backbone NH	debromo hymenaldisine Azepine CO
2.	Gate Keeper Residue	H-bond	debromo hymenaldisine Azepine NH	Glu 242 backbone CO
3.	Solvent exposed area	H-bond	Lys 191 side chain NH	debromo hymenaldisine N adjacent to carbonyl of the imidazole ring
4.	Phosphate binding region	H-bond	debromo hymenaldisine imidazole NH2	Asp 325 sidechain

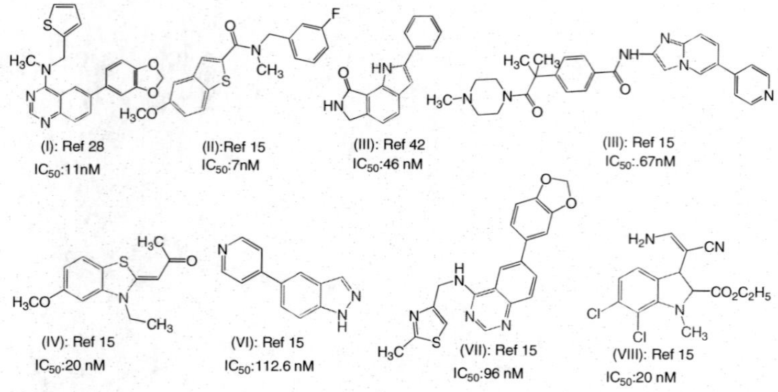

Scheme 5. Structures of some selected inhibitors of the target CLK1 with their IC_{50} values. The pharmacophore responsible for inducing the activity is coloured in blue.

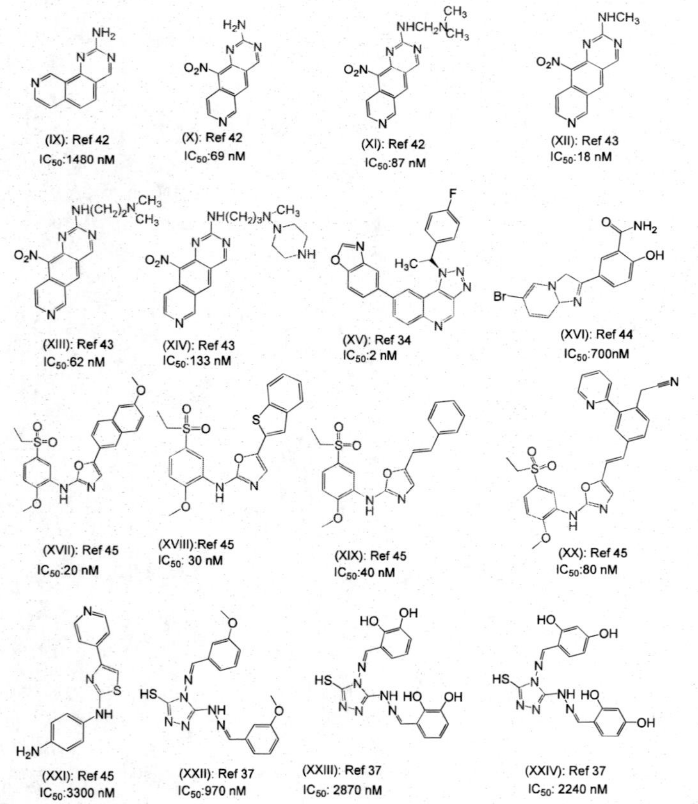

Scheme 6. Structures of few more selected inhibitors of the target CLK1 with their IC_{50} values. The pharmacophore responsible for inducing the activity is coloured in blue.

In the subsequent section, we report a series of CLK1 inhibitors that have shown considerable potency and selectivity towards this target from the last decade to till date (micro molar to nano molar).

Design of Inhibitors for the Receptor GSK3β

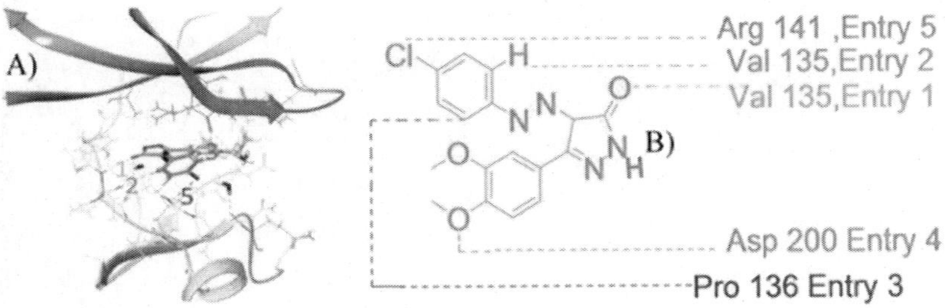

Figure 8. A; Molecular modelling 3D image of the reference compound (4E)-[(4-chlorophenyl)hydrazono]-5-(3,4-dimethoxy phenyl)-2-4 dihydro-3H-pyrazol-3-one, within the ATP binding site of GSK3β . B: Structure of the reference compound demonstrating various interactions within the residues of the active site. The water molecules have been omitted for clarity. The reference ligand is shown in green. The various interactions with different residues (shown in various colours) observed at the active site are shown in dotted lines.

Table 3. Various interactions of the reference ligand (4E)-[(4-chloro phenyl) hydrazono]-5-(3,4-dimethoxy phenyl)-2-4-di hydro 3H-pyrazol-3 one (PDB ID: 3L1S) with the receptor GSK3β

Entry	Region of interaction	Interactions involved in the active site	Residues involved	
			Donor (Source)	Acceptor
1.	Hinge region	H-bond	Val 135 backbone NH	Reference molecule pyrazole CO
2.	Hinge region	H-bond	Reference molecule meta H- of Cl benzene	Val 135 backbone CO
3.	Hinge region	H-bond	Reference molecule m-H of Cl benzene	Pro 136 backbone CO
4.	Activation loop	H-bond	Asp 200 backbone NH	Reference molecule Phe ring 3-methoxy
5.	Hydrophobic pocket	H-bond	Arg 141 side chain NH	Reference molecule Cl of Cl benzene

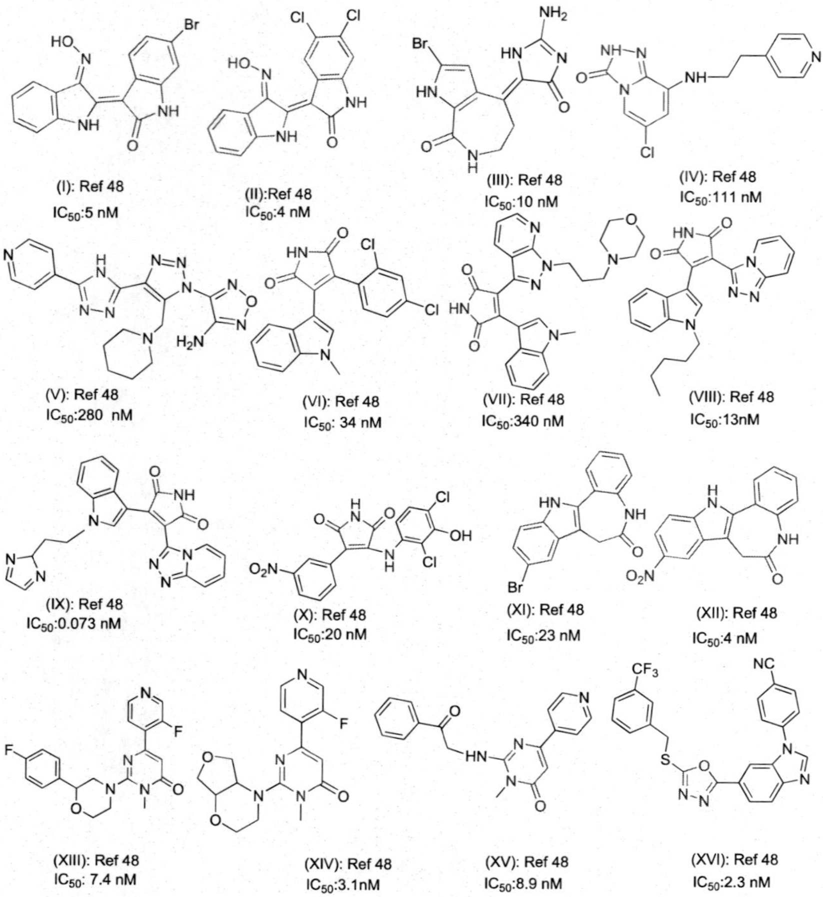

Scheme 7. Structures of some selected inhibitors of the target GSK3β with their IC_{50} values. The pharmacophore responsible for inducing the activity is coloured in blue.

To explore the possible interactions at the ATP binding site of the receptor GSK3β, herein we report the co-crystallised structure of the reference ligand (4E)-[(4-chlorophenyl)hydrazono]-5-(3,4-dimethoxy phenyl)-2-4 dihydro-3H-pyrazol-3-one incorporated in the active site using the PDB Grid: 3L1S (Figure 8) [47]. The interactions that were primarily observed are as follows: 1) A H-bonding interaction with NH of Val 135 and pyrazole carbonyl of reference molecule (Table 3, Entry 1) (Figure 8); 2) A second H-bond between Val 135 CO and CH-of Cl-benzene ring of reference ligand (Table 3, Entry 2) (Figure 8); 3) A third H- bond between H of Cl-benzene ring of reference ligand and Pro 136 (Table 3, Entry 3) (Figure 8); 4)A forth H-bond between m- Methoxy of phenyl ring of co-crystallised ligand and NH of Aspartic acid (Asp 200) (Table 3, Entry 4) (Figure 8); 5) A fifth H-bond between Cl of Cl-benzene ring of reference ligand and Arg 141 sidechain (Table 3, 5) (Figure 8).

In the section described below, we report a series of GSK3β inhibitors that have shown considerable potency and selectivity towards this target from the last decade to till date (micro molar to nano molar).

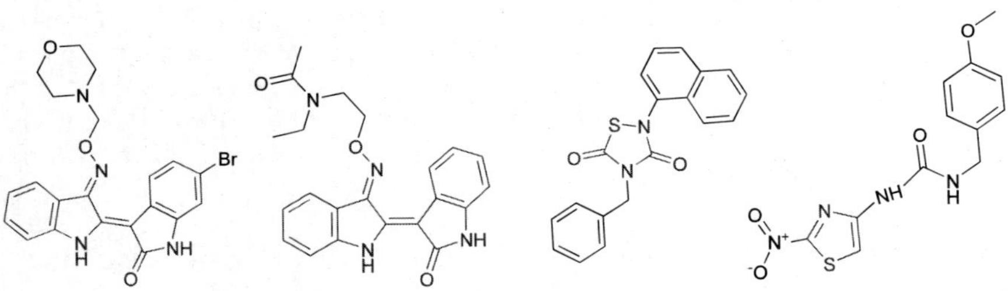

(XVII): Ref 7 IC_{50}:110 nM

(XVIII): Ref 7 IC_{50}:30 nM

(XIX): Ref 48 IC_{50}:60 nM

(XX): Ref 48 IC_{50}:100 nM

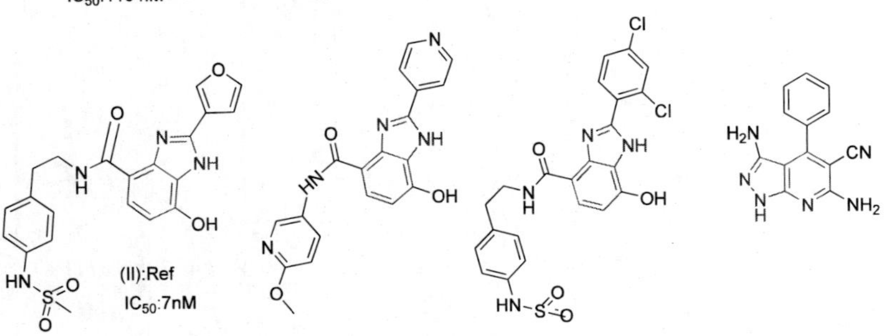

(XXI): Ref 48 IC_{50}:1nM

(XXII: Ref 48 IC_{50}: 2 nM

(XXIII): Ref48 IC_{50}:15 nM

(XXIV): Ref 48 IC_{50}:1500 nM

Scheme 8. Structures of some more selected inhibitors of the target GSK3β with their IC_{50} values. The pharmacophore responsible for inducing the activity is coloured in blue.

Design of Inhibitors for the Receptor CDK5

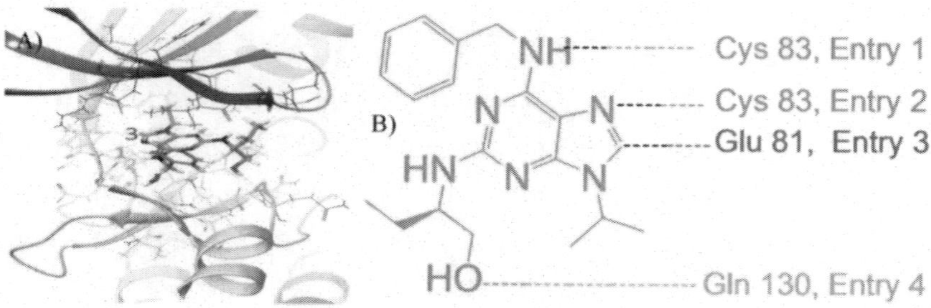

Figure 9. A; Molecular modelling 3D image of the reference compound R-Roscovitine, within the ATP binding site of CDK5; B: 2D Structure of the reference compound demonstrating various interactions within the residues of the active site. The water molecules have been omitted for clarity. The reference ligand is shown in green. The various interactions with different residues (shown in various colours) observed at the active site are shown in dotted lines.

Table 4. Various interactions of the reference ligand R-Roscovittine (PDB ID: 1UNL) with the receptor CDK5

Entry	Region of interaction	Interactions involved in the active site	Residues involved	
			Donor (Source)	Acceptor
1.	Hinge region	H-bond	Reference molecule linker NH	Cys 83 CO
2.	Hinge region	H-bond	Cys 83 NH	Reference molecule imidazole N
3.	Hinge region	H-bond	Reference molecule H between two N's of imidazole ring	Glu 81 sidechain CO
4.	Activation loop	H-bond	Reference molecule OH	Gln 130 CO

To investigate the possible interactions at the ATP binding site of the receptor CDK5, herein we report the co-crystallised structure of the reference ligand R-Roscovittine incorporated at the active site using the PDB Grid: 3UZP (Figure 8) [60]. The correspondence that were primarily observed are as follows: a) A H-bonding interaction with linker NH of

Roscovittine and Cys 83 carbonyl (Table 4, Entry 1) (Figure 9); b)A second H-bond between Cys 83 and N of imidazole ring of Roscovittine (Table 4, Entry 2) (Figure 9); c) A third H-bond between H-in between two nitrogens of imidazole and Glu 81 CO (Table 4, Entry 3) (Figure 9); d) A fourth H-bond between OH of Roscovittine and Gln130 backbone carbonyl (Table 4, Entry 4) (Figure 9).

In the subsequent section, we report a series of CDK5 inhibitors that have shown considerable potency and selectivity towards this target from the last decade to till date (micro molar to nano molar).

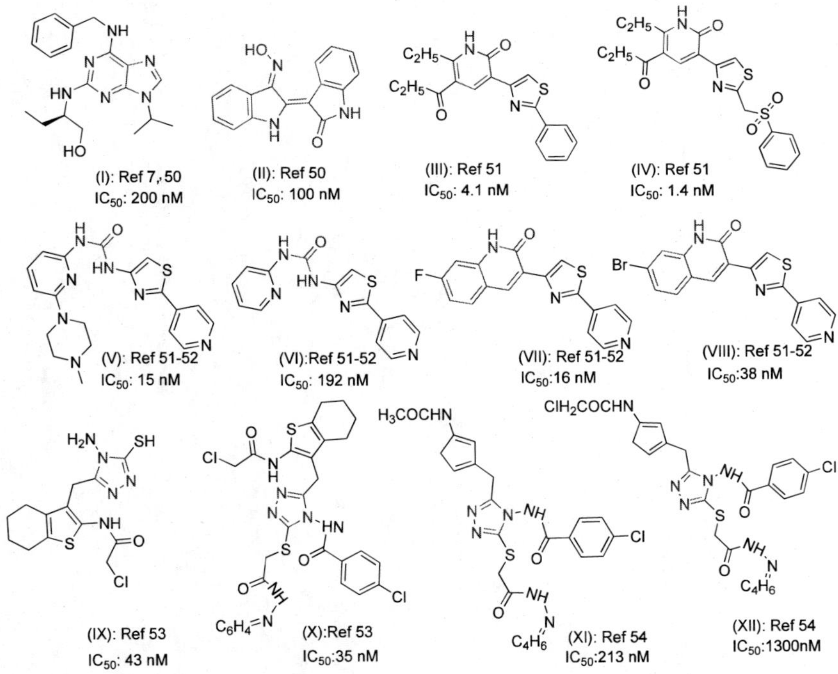

Scheme 9. Structures of some selected inhibitors of the target CDK5 with their IC_{50} values. The pharmacophore responsible for inducing the activity is coloured in blue.

Design of Inhibitors for the Receptor CK1δ

To investigate the possible interactions at the ATP binding site of the receptor CK1δ, herein we report the molecular docking studies with the reference compound (4)-[1-cyclohexyl-4-(4-fluorophenyl)-1H-imidazole-5yl] pyrimidine -2amine (PDB ID: 3UZP) with CK1δ (Figure 10) [61]. The interactions that were primarily observed are as follows: a) A H- bond between Gatekeeper Glu 83 backbone carbonyl with CH of pyrimidine ring (Entry 1, Table 5) (Figure 10); b) A second H-bonding interaction with the pyrimidine ring NH_2 and Leu 85 carbonyl (Table 5, Entry 2), (Figure 10); c) A third H-bond with the Leu NH carbonyl and pyrimidine ring N (Table 5, Entry 3), (Figure 10).

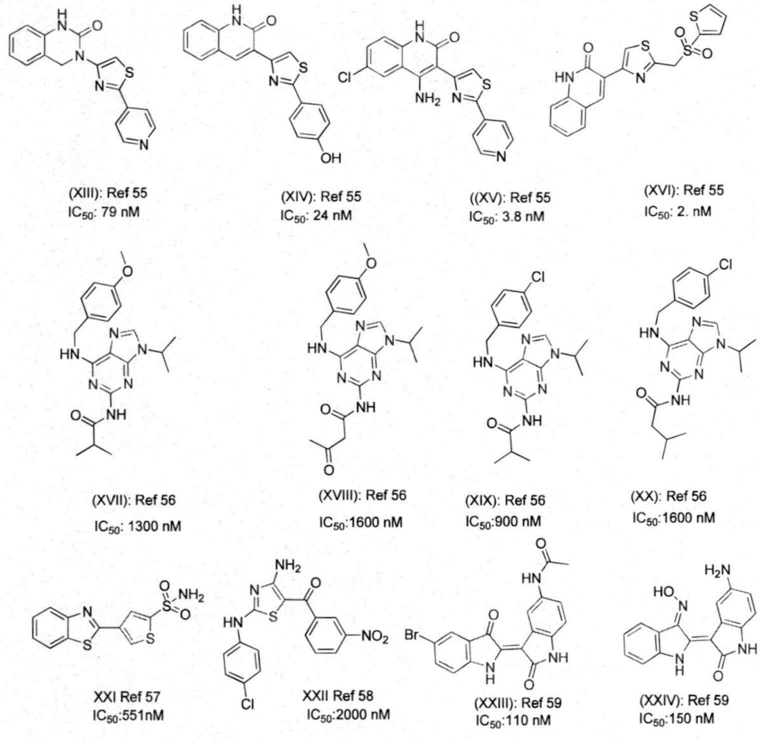

Scheme 10. Structures of some more selected inhibitors of the target CDK5 with their IC_{50} values. The pharmacophore responsible for inducing the activity is coloured in blue.

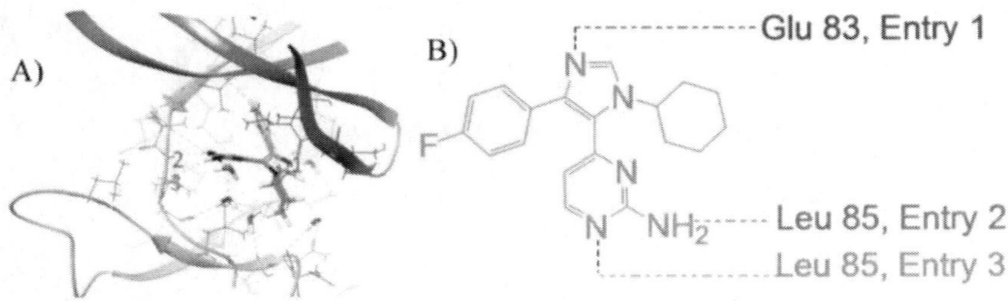

Figure 10. A; Molecular modelling 3D image of the reference compound (4)-[1-cyclohexyl-4-(4-fluorophenyl)-1H-imidazole-5yl] pyrimidine -2amine within the ATP binding site of CK1δ; B: 2D Structure of the reference compound demonstrating various interactions within the residues of the active site. The reference ligand is shown in green. The various interactions with different residues (shown in various colours) observed at the active site are shown in dotted lines.

Table 5. Various interactions of the reference ligand (4)-[1-cyclohexyl-4-(4-fluorophenyl)-1H-imidazole-5yl] pyrimidine -2amine with the receptor CK1δ

Entry	Region of interaction	Interactions involved in the active site	Residues involved	
			Donor (Source)	Acceptor
1.	Gatekeeper	H-bond	Reference molecule C-H of pyrimidine ring	Glu 83 CO
2.	Hinge region	H-bond	Reference molecule NH2 of pyrimidine	Leu 85 CO
3.	Hinge region	H-bond	Leu 85 NH	Reference molecule N of pyrimidine

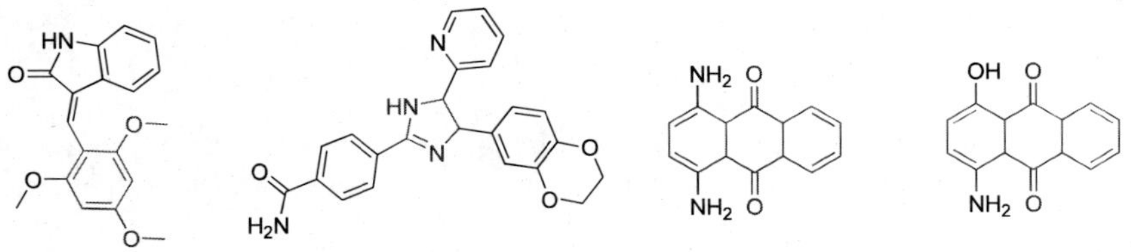

(I): Ref 61
IC_{50}:1000 nM

(II):Ref 62
IC_{50}:300 nM

(III): Ref 63
IC_{50}:330nM

(IV): Ref 63
IC_{50}:660 nM

Scheme 11. Structures of some selected inhibitors of the target CK1δ with their IC_{50} values. The pharmacophore responsible for inducing the activity is coloured in blue.

In the section described below, we report a series of CK1δ inhibitors that have shown considerable potency and selectivity towards this target from the last decade to till date (micro molar to nano molar).

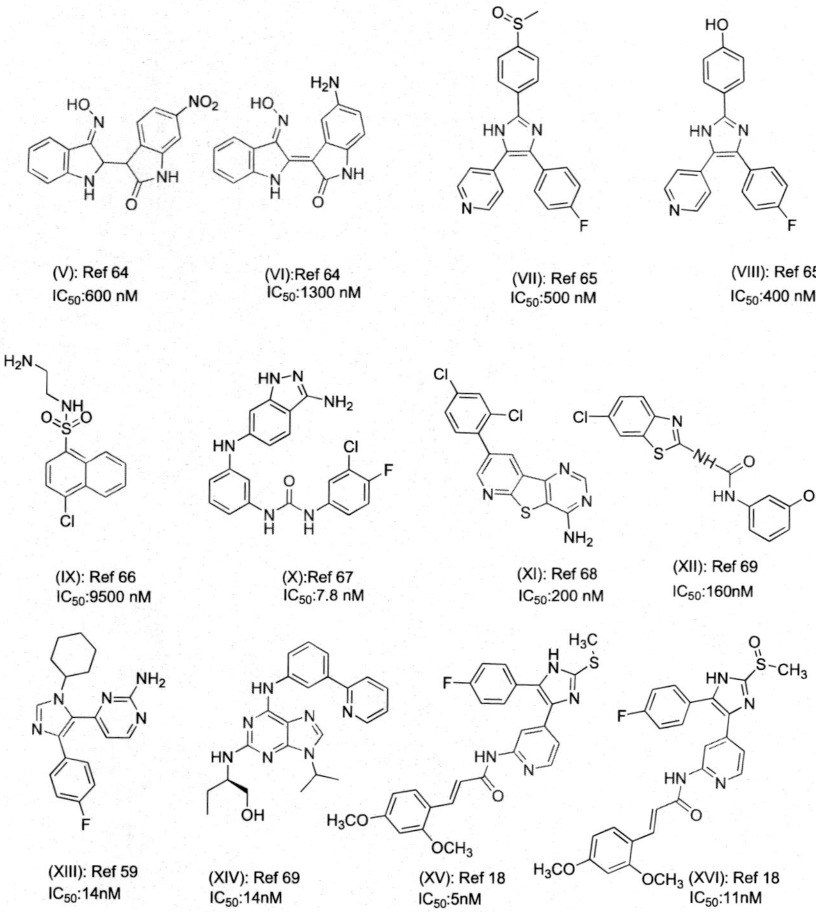

Scheme 12. Structures of some more selected inhibitors of the target CK1δ with their IC_{50} values. The pharmacophore responsible for inducing the activity is coloured in blue.

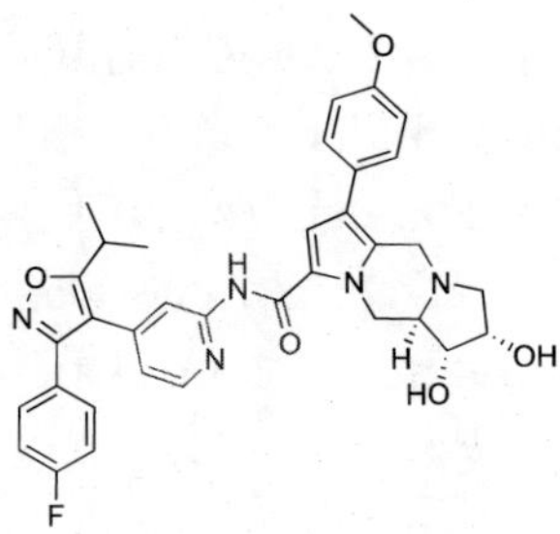

(XVII): Ref 70
IC_{50}:11 nM

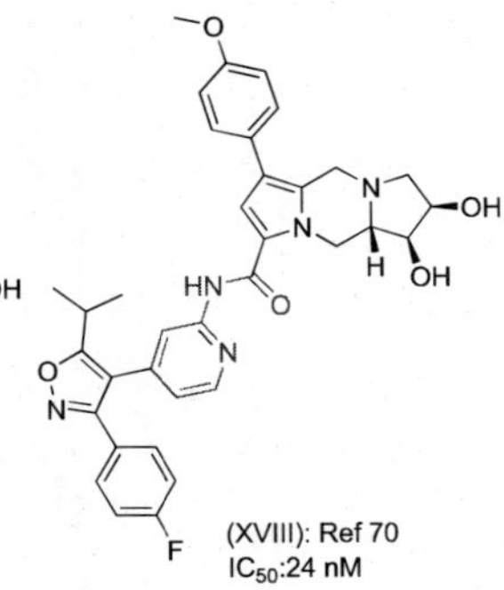

(XVIII): Ref 70
IC_{50}:24 nM

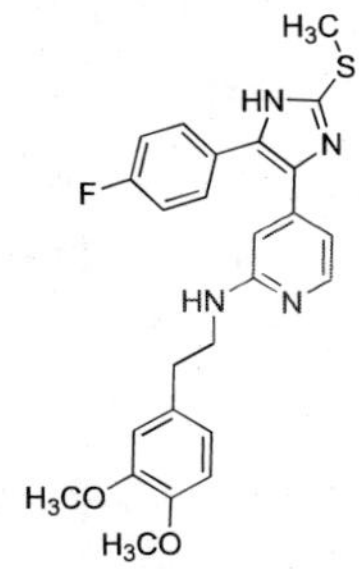

(XIX): Ref18
IC_{50}:644 nM

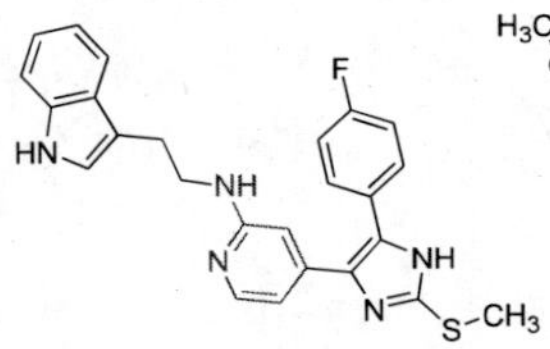

(XX):Ref 18
IC_{50}: 344 nM

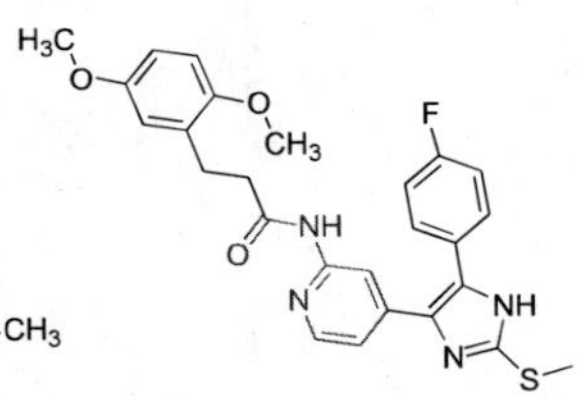

(XXI): Ref 18
IC_{50}:4nM

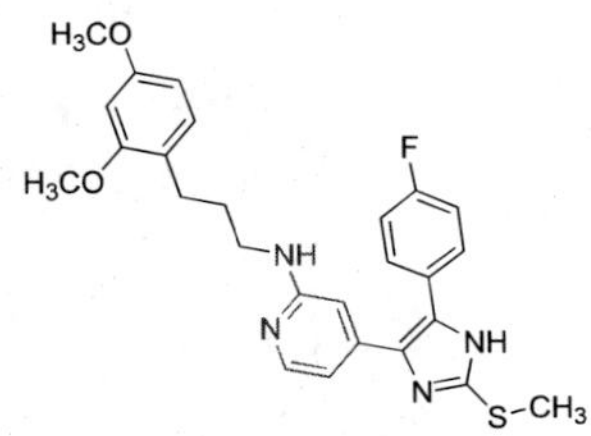

(XXII): Ref 18
IC_{50}:386nM

Scheme 13. Structures of some more selected inhibitors of the target CK1δ with their IC_{50} values. The pharmacophore responsible for inducing the activity is coloured in blue.

We postulate that inhibitors with dimensions comparable to the reference ligand at the active site of various receptors will have favourable interactions. A minor modification in the structure of the designed inhibitors could produce new interactions that may, in some cases, favor or disfavor the binding ability with the receptor, which will be simultaneously reflected in their inhibitory activities.

CONCLUSION

Statistics reveal that there has been a steep increase upto 26 million in the incidence of Alzheimer's disease amongst the elderly population in the developing countries. Additionally, it has been predicted that this data will triple by the end of the year 2050 [71]. Indeed this disorder primarily originates owing to the accumulation of intracellular deposites of microtubule associated tau proteins into neurofibrillar tangles (NFTs) at specific regions of the brains mainly driven by the hyperphosphorylation activity of the kinases. So targeting kinases to inhibit this posttranslational modification of hyperphosphorylation, represents a potential strategy for the development of Alzheimer's therapeutics, that might reduce tau aggregation and associated neuronal death. Although significant amount of work has been accomplished in this arena, we still feel that there is a lot more room to be developed in the area of better therapeutic design against Alzheimer's. Therefore through this chapter we attempt to communicate three messages that should be focused during the design of the ATP competitive inhibitors:

a) The mimetic should be designed in such a way that when it approaches the close proximity of the active site, it might enter into specific interactions with some specific residues of the active site of the ATP binding pocket, thereby inducing the inhibitory activities in the nanomolar range.
b) This can only be achieved by considering the key interactions of the reference ligand at the active site and identifying similar ligands with diversified pharmacophore that might be eligible enough to induce comparable interactions. In this way selectivity might be obtained.
c) Then by fine tuning the electron density around the identified pharmacophore by introducing/removal of suitable functionalities in appropriate locations, we might obtain better interactions with the residues of the active site of the target receptors incomparison to the co-crystallized ligand.

We feel through this strategy better understanding of Structure Activity Relationship (SAR) may be achieved, thereby fecilitating future development of more successful therapeutic agent for the treatment of neurodegenerative diseases like that of Alzheimers, superior to those of the reported ones till date.

ACKNOWLEDGMENTS

AKJ wishes to thank UGC (MANF-2015-2017-MAD-50686), New Delhi for financial support. ADK is grateful to UGC (F.4-(55)/2014(BSR)/FRP), New Delhi for financial support. The authors thank Dr. Charles R. Ashby, Jr. (St. John's University) for his editorial assistance.

Conflict of Interest

No conflict of interest was declared.

REFERENCES

[1] Prokop, A., (2013). The intricate relationship between microtubules and their associated motor proteins during axon growth and maintenance. *Neural Development*, 8 (17):1-10.

[2] Bodea, L-G., Eckert, A., Ittner, L. M., Piguet, O. and Götz, J. (2016). Tau physiology and pathomechanisms in frontotemporal lobar degeneration. *J. Neurochem.*, 138 (71): 1-24.

[3] Smith, D. S., Järlfors U. and Cameron B. F., (1975). Morphological evidence for the participation of microtubules in axonal transport. *Ann N Y Acad Sci*, 253: 472-506.

[4] Fichou, Y., Al-Hilaly, Y. K., Devred, F., Smet-Nocca, C., Tsyetkoy, P. O., Verelist, J., Winderickx, J., Geukens, N., Vanmechelen, E., Perrotin, A., Serpell, L., Hanseeuw, B. J., Medina, M., Buee L. and Landrieu, I. (2019). The elusive tau molecular structures: can we translate the recent breakthroughs into new targets for intervention? *Acta Neuropathol Commun.*, 7 (31):1-17.

[5] Johnson, G. V. and Stoothoff, W. H. (2004). Tau phosphorylation in neuronal cell function and dysfunction. *J Cell Sci.*, 117: 5721-5729.

[6] Verwilst, P., Kim, H. S., Kim, S., Kang C. and Kim, J. S. (2018). Shedding light on tau protein aggregation: the progress in developing highly selective fluorophores. *Chem. Soc. Rev.*, 47(7): 2249-2265.

[7] Martin, L., Latypova, X., Wilson, C. M., Magnaudeix, A., Perrin, M. L., Yardin, C. and Terro, F. (2013). Tau protein kinases: involvement in Alzheimer's disease. *Ageing Res. Rev.*, 12: 289-309.

[8] Phillip, A. S. and Murray, W. (2011). Protein kinase biochemistry and drug discovery. *Bioorganic Chemistry*, 39: 192-210.

[9] Weber, T. J. (2010). Protein Kinases. *Comprehensive Toxicology*, 2: 473-493.

[10] Hanks, S. K. and Quinn, A. M. (1991). Protein kinase catalytic domain sequence database: identification of conserved features of primary structure and classification of family members. *Methods in Enzymology*, 200: 38-62.

[11] Hunter, T. (1991). Protein kinase classification. *Methods in Enzymology*, 200: 3-37.

[12] Manning, G., Whyte, D. B., Martinez, R., Hunter, T. and Sudarsanam, S. (2002). The protein kinase complement of the human genome. *SCIENCE*, 298: 1912-1934.

[13] Wang, D., Wang, F., Tan, Y., Dong, L., Chen, L., Zhu, W. and Wang, H. (2012). Discovery of potent small molecule inhibitors of DYRK1A by structure-based virtual screening and bioassay. *Bioorg. Med. Chem. Lett.*, 22: 168-171.

[14] Berman, H. M., Westbrook, J., Feng, Z., Gilliland, G., Bhat, T. N., Weissig, H., Shindyalov I. N. and Bourne, P. E. (2000). The Protein Data Bank, *Nucleic Acids Res.*, 28: 235-242.

[15] Walter, A., Chaikuad, A., Helmer, R., Loaëc, N., Preu, L., Ott, I., Knapp, S., Meijer L. and Kunick, C. (2018). Molecular structures of cdc2-like kinases in complex with a new inhibitor chemotype. *PLoS One*, 13: 1-29. e0196761.

[16] Iqbal K. and Grundke-Iqbal, I. (1995). Alzheimer abnormally phosphorylated tau is more hyperphosphorylated than the fetal tau and causes the disruption of microtubules. *Neurobiology of Aging*, 16 (3): 375-379.

[17] Engmann O. and Giese, K. P. (2009). Crosstalk between Cdk5 and GSK3β: Implications for Alzheimer's Disease. *Frontiers in Molecular Neurosciences*. 2: 1-5.

[18] Halekotte, J., Witt, L., Ianes, C., Krüger, M., Bührmann, M., Rauh, D., Pichlo, C., Brunstein, E., Luxenburger, A., Baumann, U., Knippschild, U., Bischof, J. and Peifer, C. (2017). Optimized 4,5-diarylimidazoles as potent/selective inhibitors of protein kinase CK1δ and their structural relation to p38α MAPK. *Molecules*, 22: 522-559.

[19] Johnson, L. N., Lowe, E. D., Noble, M. E. M. and Owen, D. J. (1998). The structural basis for substrate recognition and control by protein kinases. *FEBS Letters*, 430: 1-11.

[20] Kornev, A. P. and Taylor, S. S. (2010). Defining the conserved internal architecture of a protein kinase., Biochimica et Biophysica Acta (BBA) - *Proteins and Proteomics*, 1804 (3): 440-444.

[21] Ogawa, Y., Nonaka, Y., Goto, T., Ohnishi, E., Hiramatsu, T., Kii, I., Yoshida, M., Ikura, T., Onogi, H., Shibuya, H., Hosoya, T., Ito N. and Hagiwara M. (2010). Development of a novel selective inhibitor of the Down syndrome-related kinase Dyrk1A. *Nature Communications*, 1(86): 1-9.

[22] Jarhad, D. B., Mashelkar, K. K., Kim, H. R., Noh, M. and Jeong, L. S. (2018). Dual-Specificity Tyrosine Phosphorylation-Regulated Kinase 1A (DYRK1A) Inhibitors as Potential Therapeutics. *Journal of Medicinal Chemistry*, 61(22): 9791-9810.

[23] Debdab, M., Carreaux, F., Renault, S. Soundararajan, M., Fedorov, O., Filippakopoulos, P., Lozach, O. Babault, L., Tahtouh, T., Baratte, B., Ogawa, Y., Hagiwara, M., Eisenreich, A., Rauch, U., Knapp, S., Meijer, L. and Bazureau, J. P., (2011). Leucettines, a Class of Potent Inhibitors of cdc2-Like Kinases and Dual Specificity, Tyrosine Phosphorylation Regulated Kinases Derived from the Marine Sponge Leucettamine B: Modulation of Alternative Pre-RNA Splicing *Journal of Medicinal Chemistry*, 54: 4172-4186.

[24] Burgy, G., Limanton, E., Carreaux, F., Durieu, E., Meijer, L., and Bazureau, J.P. 2014.Exploring the Synthesis of New 1-(4-Substitutedphenylamino) imidazo[1,5-a] indol-3-one Derivatives as

Cyclized Analogs of Leucettines. *Journal of Applied Pharmaceutical Science*, 4: 25-32.

[25] Neagoie, C., Vedrenne,E., Buron, F., Merour, J.Y., Rosca, S., Bourg, S., Lozach, O., Meijer, L., Baldeyrou, B., Lansiaux, A. and Routier, S. (2012). Synthesis of chromeno[3,4-b] indoles as Lamellarin D analogues: A novel DYRK1A inhibitor class. *European Journal of Medicinal Chemistry*, 49: 379-396.

[26] Echalier, A., Bettayeb, K., Ferandin, Y., Lozach, O., Clement, M., Valette, A., Liger, F., Marquet, B., Morris, J.C., Endicott, J.A., Joseph, B. and Meijer, L. (2008). Meriolins (3-(pyrimidin-4-yl)-7-azaindoles): synthesis, kinase inhibitory activity, cellular effects, and structure of a cdk2/cyclin a/meriolin complex. *J Med Chem*, 51: 737-751.

[27] Mott, B.T., Tanega, C., Shen, M., Maloney, D.J., Shinn, P., Leister, W., Marugan, J.J., Inglese, J., Austin, C.P., Misteli, T., Auld, D.S. and Thomas, C.J. (2009). Evaluation of substituted 6-arylquinazolin-4-amines as potent and selective inhibitors of cdc2-like kinases (Clk) *Bioorganic Medicinal Chemistry Letters*,19: 6700-6705.

[28] Rosenthal, A.S., Tanega, C., Shen, M., Mott, B.T., Bougie, J.M., Nguyen, D.T., Misteli, T., Auld, D.S., Maloney, D. J. and Thomas, C.J. (2011). Potent and selective small molecule inhibitors of specific isoforms of Cdc2-like kinases (Clk) and dual specificity tyrosine-phosphorylation-regulated kinases (Dyrk). *Bioorganic Medicinal Chemistry Letters*, 21: 3152-3158.

[29] Demange, L., Abdellah, F.N., Lozach, O., Ferandin, Y., Gresh, N., Meijer, L. and Galons, H. (2013). Potent inhibitors of CDK5 derived from roscovitine: Synthesis, biological evaluation and molecular modelling. *Bioorganic Medicinal Chemistry Letters*, 23: 125-131.

[30] Ahmadu, A., Abdulkarim, A., Grougnet,R., Myrianthopoulos, V., Tillequin, F., Magiatis, P. and Skaltsounis, A. L. (2010). Two new peltogynoids from acacia nilotica delile with kinase inhibitory activity. *Planta Med*, 76: 458-460.

[31] Foucourt, A., Hedou, D., Dubouilh-Benard, C., Girard, A., Taverne, T., Casagrande, A. S., Desire, L., Leblond B. and Besson, T. (2014).

Design and Synthesis of Thiazolo[5,4-f] quinazolines as DYRK1A Inhibitors, Part II. *Molecules*, 19: 15411 – 15439.

[32] Shengule, S. R., Loa-Kum-Cheung, W. L., Parish, C. R., Blairvacq, M., Meijer, L., Nakao, Y. and Karuso, P. (2011). A One-Pot Synthesis and Biological Activity of Ageladine A and Analogues. *Journal of Medicinal Chemistry*, 54: 2492-2503.

[33] Sanchez, C., Salas, A. P., Brana, A.F., Palomino, M., Pineda-Lucena, A., Carbajo, R. J., Mendez, C., Moris, F. and Salas, J.A. (2009). Generation of potent and selective kinase inhibitors by combinatorial biosynthesis of glycosylated indolocarbazoles. *Chemical Communications*, 4118-4120.

[34] Sun, Q-Z., Lin, G-F., Li, L-L., Jin, X-T., Huang, L-Y., Zhang, G., Yang, W., Chen, K., Xiang, R., Chen, C., Wei, Y-Q., Lu, G-W and Yang, S-Y (2017). Discovery of Potent and Selective Inhibitors of Cdc2-Like Kinase 1 (CLK1) as a New Class of Autophagy Inducers. *Journal of Medicinal Chemistry*, 60: 6337-6352.

[35] Myrianthopoulos, V., Kritsanida, M., Gaboriaud-Kolar, N., Magiatis, P., Ferandin, Y., Durieu, E., Lozach, O., Cappel, D., Soundararajan, M., Filippakopoulos, P., Sherman, W., Knapp, S., Meijer, L., Mikros and E., Skaltsounis, L. (2013). Novel Inverse Binding Mode of Indirubin Derivatives Yields Improved Selectivity for DYRK Kinases, ACS *Medicinal Chemistry Letters*, 4: 22-26.

[36] Falke, H., Chaikuad, A., Becker, A., Loaëc, N., Lozach, O., Jhaisha, S. A., Becker, W., Jones, P. G., Preu, L., Baumann, K., Knapp, S., Meijer L. and Kunick, C., (2015). 10-Iodo-11H-indolo[3,2-c] quinoline-6-carboxylic Acids Are Selective Inhibitors of DYRK1A. *Journal of Medicinal Chemistry*, 58: 3131-3143.

[37] Jain, A. K., Karthikeyan, C., McIntosh, K. D., Tiwari, A. K., Trivedi, P. and Dutt Konar A. (2019). Unravelling the potency of 4,5-diamino-4H-1,2,4 triazole-3-thiol derivatives for kinase inhibition using a rational approach. *New Journal of Chemistry*, 43: 1202-1215.

[38] Anderson, K., Chen, Y., Chen, Z., Dominique, R., Glenn, K., He, Y., Janson, C., Luk, K.C., Lukacs, C., Polonskaia, A., Qiao, Q., Railkar, A., Rossman, P., Sun, H., Xiang, Q., Vilenchik, M., Wovkulich, P.,

Zhang, X. (2013). Pyrido[2,3-d] pyrimidines: Discovery and preliminary SAR of a novel series of DYRK1B and DYRK1A inhibitors. *Bioorganic Medicinal Chemistry Letters*, 23: 6610-6615.

[39] Yoshida, K., Itoyama, R., Yamahira, M., Tanaka, J., Loaec, N., Lozach, O., Durieu, E., Fukuda, T., Ishibashi, F., Meijer, L. and Iwao, M., (2013). Synthesis, resolution, and biological evaluation of atropisomeric (aR)- and (aS)-16-methyllamellarins n: unique effects of the axial chirality on the selectivity of protein kinases inhibition. *Journal of Medicinal Chemistry*, 56: 7289-7301.

[40] Leblond, B., Casagrande, A.-S., Desire, L., Foucourt, A. and Besson, T. (2016). *DYRK1 inhibitors and uses thereof*. Patent No., US 9,446,044 B2.

[41] Bullock, A. N., Das, S., Debreczeni, J. E., Rellos, P., Fedorov, O., Niesen, F. H., Guo, K., Papagrigoriou, E., Amos, A. L., Cho, V., Turk, B. E., Ghosh, G., and Knapp, S. (2009). Kinase Domain Insertions Define Distinct Roles of CLK Kinases in SR Protein Phosphorylation. *Structure*, 17: 352-362.

[42] Zeinyeha, W., Esvana, Y.J., Josselinb, B., Baratteb, B., Bachb, S., Nautona, L., Thérya, V., Ruchaudb, S., Anizona, F., Giraud, F., and Moreaua, P., (2019). Kinase inhibitions in pyrido[4,3-h] and [3,4-g] quinazolines: Synthesis, SAR and molecular modeling studies. *Bioorganic and Medicinal Chemistry*. doi.org/10.1016/j.bmc.2019.04.005.

[43] Tazarki, H., Zeinyeh, W., Esvan, Y. J., Knapp, S., Chatterjee, D., Schröder, M., Joerger, A. C., Khiari, J., Josselin, B., Baratte, B., Bach, S., Ruchaud, S., Anizon, F., Giraud F. and Moreau, P. (2019). New pyrido[3,4-g] quinazoline derivatives as CLK1 and DYRK1A inhibitors: synthesis, biological evaluation and binding mode analysis. *European Journal of Medicinal Chemistry*, doi.org/10.1016/j.ejmech.2019.01.052.

[44] Lawson, M., Rodrigo, J., Baratte, B., Robert, T., Delehouze, C., Lozach, O., Ruchaud, S., Bach, S., Brion, J-D Alami, M.,and Hamze, A., (2016). Synthesis, biological evaluation and molecular modeling

studies of imidazo[1,2-a] pyridines derivatives as protein kinase inhibitors. *European Journal of Medicinal Chemistry*, 123: 105-114.

[45] Murar, M., Dobias, J., Sramel, P., Addova, G., Hanquet G., and Boha, A. (2017). Novel CLK1 inhibitors based on N-aryloxazol-2-amine skeleton - A possible way to dual VEGFR2 TK/CLK ligands. *European Journal of Medicinal Chemistry*, 126: 754-761.

[46] Walter, A., Chaikuad, A., Loaëc, N., Preu, L., Knapp, S., Meijer, L., Kunick, C., and Koch, O. (2017). Kernel Multitask Regression for Toxicogenetics. *Molecular Informatics*, 36: 1600123-1600134.

[47] Arnost, M., Pierce, A., Haar, E. T., Lauffer, D., Madden, J.,Tanner K. and Green, J. (2010). 3-Aryl-4-(arylhydrazono)-1H-pyrazol-5-ones: Highly ligand efficient and potent inhibitors of GSK3β. *Bioorg. Med. Chem. Lett.*, 20: 1661-1664.

[48] Khan, I., Tantray, M. A., Alam M. S. and Hamid H. (2017). Natural and synthetic bioactive inhibitors of glycogen synthase kinase. *European Journal of Medicinal Chemistry*, 125: 464 - 477.

[49] Mapelli, M., Massimiliano, L., Crovace, C., Seeliger, M. A., Tsai, L.H., Meijer L. and Musacchio, A. (2005). Mechanism of CDK5/p25 Binding by CDK Inhibitors. *Journal of Medicinal Chemistry*, 48: 671-679.

[50] Rzasa, R. M., Kaller, M. R., Liu, G., Magal, E., Nguyen, T. T., Osslund, T. D., Powers, D., Santora, V. J., Viswanadhan, V.N., Wang, H. L., Xiong, X., Zhong W. and Norman, M. H., (2007). Structure–activity relationships of 3,4-dihydro-1H-quinazolin-2-one derivatives as potential CDK5 inhibitors. *Bioorganic and Medicinal Chemistry*, 15: 6574 - 6595.

[51] Kaller, M. R., Zhong, W., Henley, C., Magal, E., Nguyen, T. T., Powers, D., Rzasa, R. M., Wang, W., Xiong, X. and Norman, M. H., (2009). Design and synthesis of 6-oxo-1,6-dihydropyridines as CDK5 inhibitors. *Bioorganic Medicinal Chemistry Letters*, 19: 6591-6594.

[52] Shiradkar, M. R., Thomas, J., Kanase V. and Dighe, R. (2011). Studying synergism of methyl linked cyclohexyl thiophenes with triazole: Synthesis and their cdk5/p25 inhibition activity.European *Journal of Medicinal Chemistry*, 46:2066 - 2074.

[53] Shiradkar, M. R., Akula, K. C., Dasari, V., Baru, V., Chiningiri, B., Gandhi S. and Kaur, R. (2007). Clubbed thiazoles by MAOS: A novel approach to cyclin-dependent kinase 5/p25 inhibitors as a potential treatment for Alzheimer's disease. *Bioorganic and Medicinal Chemistry*, 15: 2601-2610.

[54] Zhong, W., Liu, H., Kaller, M. R., Henley, C., Magal, E., Nguyen, T., Osslund, T. D., Powers, D., Rzasa, R. M., Wang, H. L., Wang, W., Xiong, X., Zhang J. and Norman, M. H. (2007). Design and synthesis of quinolin-2(1H)-one derivatives as potent CDK5 inhibitors. *Bioorganic & Medicinal Chemistry Letters*, 17: 5384-5389.

[55] Vandromme, L., Legraverend, M., Kreimerman, S., Lozach,O., Meijer L. and Grierson, D. S. (2007). Synthesis and structure-activity relationship of 4-(1, 3-benzothiazol-2-yl)-thiophene-2-sulfonamides as cyclin-dependent kinase 5 (cdk5)/p25 inhibitors. *Bioorganic & Medicinal Chemistry*, 15:130-141.

[56] Malmström, J., Viklund, J., Slivo, C., Costa, A., Maudet, M., Sandelin, C., Hiller, G., Olsson, L. L., Aagaard, A., Geschwindner, S., Xue Y. and Vasänge, M. (2012). *Bioorganic & Medicinal Chemistry Letters*, 22: 5919–5923.

[57] Laha, J. K., Zhang, X., Qiao, L., Liu, M., Chatterjee, S., Robinson, S., Kosik, K. S. and Cuny G. D. (2011). Synthesis of novel 5-substituted indirubins as protein kinases inhibitors. *Bioorganic & Medicinal Chemistry Letters*, 21: 2098–2101.

[58] Beauchard, A., Ferandin, Y., Frere, S., Lozach, O., Blairvacq, M., Meijer, L., Thiery, V. and Besson, T. (2006). *Bioorganic & Medicinal Chemistry*, 14:6434-6443.

[59] Long, A., Zhao, H. and Huang, X., (2012). Structural basis for the interaction between casein kinase 1 delta and a potent and selective inhibitor. *Journal of Medicinal Chemistry*, 2012, 55: 956 - 960.

[60] Mashhoon, N., DeMaggio, A. J., Tereshko, V., Bergmeieri, S. C., Egli, M., Hoekstra, M. F. and Kure, J. (2000). Crystal Structure of a Conformation-selective Casein Kinase-1 Inhibitor. *Journal of Biology and Chemistry*., 275: 20052-20060.

[61] Rena, G., Bain, J., Elliott M. and Cohen, P. (2004). D4476, a cell-permeant inhibitor of CK1, suppresses the site-specific phosphorylation and nuclear exclusion of FOXO1a. *EMBO Rep*. 5:60-65.

[62] Cozza, G., Gianoncelli, A., Montopoli, M., Caparrotta, L., Venerando, A., Meggio, F., A. Pinna, L. A., Zagotto, G. and Moro, S. (2008). Identification of novel protein kinase CK1 delta (CK1δ) inhibitors through structure-based virtual screening. *Bioorganic & Medicinal Chemistry Letters*, 18: 5672-5675.

[63] Beauchard, A., Laborie, H., Rouillard, H., Lozach, O., Ferandin, Y., Guevel, R. L., Guillouzo, C. G., Meijer, L., Besson T. and Thiery, V., (2009). Synthesis and kinase inhibitory activity of novel substituted indigoids. *Bioorganic & Medicinal Chemistry*, 17: 6257-6263.

[64] Shanware, N. P., Williams, L. M., Bowler, M. J. and Tibbetts R. S. (2009). Non-specific in vivo inhibition of CK1 by the pyridinyl imidazole p38 inhibitors SB 203580 and SB 202190. *BMB Rep*. 42: 142-147.

[65] Chijiwa, T., Hagiwara, M. and Hidaka, H., (1989). A newly synthesized selective casein kinase I inhibitor, N-(2-aminoethyl)-5-chloroisoquinoline-8-sulfonamide, and affinity purification of casein kinase I from bovine testis. *Journal of Biological Chemistry*, 264: 4924 -4927.

[66] Yang, L. L. and Lee, G. B., (2012). Discovery of N6-phenyl-1H-pyrazolo[3,4-d] pyrimidine-3,6-diamine derivatives as novel CK1 inhibitors using common-feature pharmacophore model based virtual screening and hit-to-lead optimization. *European Journal of Medicinal Chemistry*, 56: 30-38.

[67] Loidreau, Y., Deau, E., Marchand, P., Nourrisson, M. R., Loge, C., Coadou, G., Loaec, N., Meijer, L. and Besson, T. (2015). Synthesis and molecular modelling studies of 8-arylpyrido [3′, 2′: 4, 5] thieno [3, 2-d] pyrimidin-4-amines as multitarget Ser/Thr kinases inhibitors. *European Journal of Medicinal Chemistry*, 92: 124-134.

[68] Benek, O., Hroch, L., Aitken, L., Moore, F. G., Vinklarova, L., Kuca, K., Perez, D. I., Perez, C., Martinez, A., Fisar Z. and Musilek, K.

(2018). 1-(Benzo[d]thiazol-2-yl)-3-phenylureas as dual inhibitors of casein kinase 1 and ABAD enzymes for treatment of neurodegenerative disorders. *Journal of Enzyme Inhibition and Medicinal Chemistry*, 33: 665 -670.

[69] Oumata, N., Bettayeb, K., Ferandin, Y., Demange, L., Giral, A.L.,Goddard, M. L., Myrianthopoulos, V., Mikros, E., Flajolet, M., Greengard, P., Meijer, L. and Galons, H., (2008). Roscovitine-derived, dual-specificity inhibitors of cyclin-dependent kinases and casein kinases 1. *Journal of Medicinal Chemistry*, 51: 5229-5242.

[70] Luxenburger, A., Schmidt, D., Ianes, C., Pichlo, C., Krüger, M., Drathen, T. V., Brunstein, E., Gainsford, G.J., Baumann, U., Knippschild U. and Peifer, C., (2019). Design, synthesis and biological evaluation of isoxazole-based ck1 inhibitors modified with chiral pyrrolidine scaffolds. *Molecules*, 24: 873-907.

[71] Jain, P., Karthikeyan, C., Moorthy, N. S. H. N., Waiker, D. K., Jain, A. and Trivedi, P. (2014). Human CDC2-Like Kinase 1 (CLK1): A Novel Target for Alzheimer's Disease. *Current Drug Targets*, 15: 539-550.

INDEX

D

I

K

L

M

N

O

P

R

S

Related Nova Publications

UNDERSTANDING HALOGENATION

EDITOR: Vladimir T. Phelps

SERIES: Chemistry Research and Applications

BOOK DESCRIPTION: *Understanding Halogenation* explores how halogenation stress could participate in neuronal damage of the nervous system and parkinsonian deficits.

SOFTCOVER ISBN: 978-1-53615-947-9
RETAIL PRICE: $82

IMINES: AN OVERVIEW

EDITOR: Gordon Sjögren

SERIES: Chemistry Research and Applications

BOOK DESCRIPTION: In the opening chapter of *Imines: An Overview,* attempts will be made to provide a brief survey of di-imines derived from benzil hydrazones through imine chemistry. Imines are compounds containing an azomethine linkage C=N connected to hydrogen or carbon atoms.

SOFTCOVER ISBN: 978-1-53616-007-9
RETAIL PRICE: $95

To see a complete list of Nova publications, please visit our website at www.novapublishers.com

Related Nova Publications

Crystal Violet: Production, Applications and Precautions

Editor: Victor Duffet

Series: Chemistry Research and Applications

Book Description: *Crystal Violet: Production, Applications and Precautions* opens by presenting the main factors influencing the metachromatic phenomenon. The hypsochromic effect is due to the symmetry of the molecule providing the common electronic signal around the central carbon atom.

Softcover ISBN: 978-1-53615-806-9
Retail Price: $82

Supercritical Carbon Dioxide

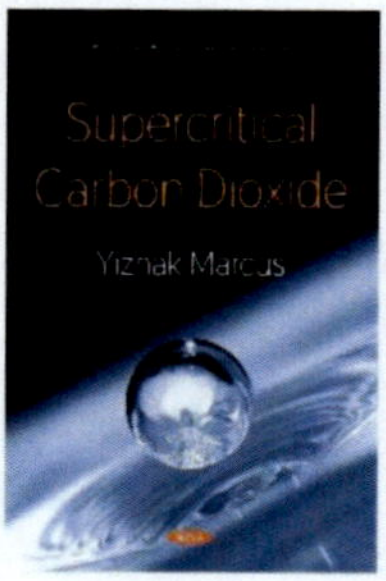

Author: Yizhak Marcus

Series: Chemistry Research and Applications

Book Description: Co-solvents are frequently added to SCD, and the properties of the resulting supercritical fluid mixtures are described in this book. Once either pure SCD or its mixture with a co-solvent have been produced, and the temperature and pressure have been suitably adjusted, the solvent power of this fluid is employed in a great variety of applications that are reviewed and discussed. In particular, the manner of adherence of such fluids to the requirements from 'green' solvents are assessed.

Softcover ISBN: 978-1-53615-165-7
Retail Price: $82

To see a complete list of Nova publications, please visit our website at www.novapublishers.com